THE PREMIXING METHOD
Principle, Design and Construction

The Premixing Method

Principle, Design and Construction

Edited by
COASTAL DEVELOPMENT INSTITUTE OF
TECHNOLOGY (CDIT), Tokyo, Japan

A.A. BALKEMA PUBLISHERS / LISSE / ABINGDON / EXTON (PA) / TOKYO

Library of Congress Cataloging-in-Publication Data

The premixing method : principle, design, and construction / edited by Coastal
  Development Institute of Technology (CDIT), Japan.
      p. cm.
    Includes index.
    ISBN 90-5809-547-9
      1. Soil mechanics. 2. Foundations. I. Engan Kaihatsu Gijutsu Kenkyu Senta (Japan)

TA710.P695 2003
624.1'51363--dc21

2003050202

Printed by: Grafisch Produktiebedrijf Gorter, Steenwijk, The Netherlands.

Published by: A.A. Balkema Publishers, a member of Swets & Zeitlinger Publishers
www.balkema.nl and www.szp.swets.nl

ISBN 90 5809 547 9

Contents

Preface

The liquefaction of soil deposits may seriously affect the structures built on those deposits, and so a critical part of the design of structures is an assessment of the likelihood of liquefaction. If the assessment reveals that liquefaction could occur, countermeasures are usually applied to the deposits.

The premix method described in this manual, in which soil used for reclamation is first treated by adding a small amount of cement, was developed to prevent liquefaction. The treated soil will acquire cohesive strength due to the chemical reaction of the cement in the water and change to a non-liquefying material. The premix method thus reduces the whole construction period, because the countermeasure work is conducted together with the reclamation process. The method has additional merits including enhancement of bearing capacity and reduction of earth pressure due to the increase in cohesion.

The premix method was developed from 1985 and was used for the first time for the Tokyo Bay Aqua-Line project in 1990. Since then, it has been used in several projects such as the restoration following the 1995 Hyogo-ken Nambu Earthquake disaster. In 1998, the Coastal Development Institute of Technology in Japan published technical guidelines for the premix method based on these two case examples for Japanese engineers.

This manual is the English version of those guidelines. I hope that it will serve as a useful reference for engineers involved in geotechnical problems around the world.

List of Committee Members

Yamazaki,Hiroyuki   Port and Airport Research Institute

Katsuumi,Tsutomu   Coastal Development Institute of Technology

Satou,Shigeki   Coastal Development Institute of Technology

Fujiwara,Toshimitsu   Premixing Method Association

Miura,Hitoshi   Premixing Method Association

Ninomiya,Koji   Premixing Method Association

Toriihara,Makoto   Premixing Method Association

Yoshida,Takaaki   Premixing Method Association

List of Technical Terms

Reclamation	formation of ground by reclamation, backfill, fill etc.
Reclamation soil	soil used to form ground by reclamation, it refers to sand, sandy soil, gravel, and other coarse grain soils, but excludes clay and other fine grain soils.
Stabilizer	cement type stabilizing material that is added to and mixed with reclamation soil to improve its properties.
Percentage of stabilizer added	percentage of stabilizer added represents the percentage by mass of the stabilizer relative to the dry soil.
Separation inhibitor	water soluble high polymer material that prevents the separation of the soil particles and the stabilizer in water.
Quantity of Separation inhibitor added	quantity of separation inhibitor added represents the ratio by mass of the agent (mg) to the dry soil (kg).
Treated soil	soil that has been treated by the addition of the stabilizer.
Untreated soil	soil in untreated condition to which stabilizer has not been added.
Treated ground	ground formed using treated soil.
Untreated ground	ground formed with untreated soil.
Unconfined compressive strength	the maximum deviator stress when a cylindrical specimen of the treated soil undergoes compressive failure in the axial direction, represented by (q_{ud}).
Cyclic shear stress ratio	the value obtained by dividing the cyclic shear stress (τ_d) by the initial effective consolidation stress (σ_c'), represented as τ_d/σ_c'.

Cyclic strength   cyclic shear stress ratio τ_d/σ_c' which causes failure at time of the 20th cycle (N_f) until failure.

Laboratory mix proportion   mix proportion that clarifies the relationship between the strength and factors such as the type and percentage added of the stabilizer and the type and percentage added of the separation inhibitor with the strength.

Mix proportion at job site   mix proportion used to execute the method on site.

Design strength   strength used for design; its index is the unconfined compressive strength (q_{udd}).

Laboratory strength   mean value of the strength obtained by the laboratory mix proportion test (q_{udl}).

Field strength   mean value of the strength actually obtained at a site (q_{udf}).

Required additional rate   ratio of the laboratory strength to the field strength obtained by the same mix proportions (α).

Modified strength for   product of the design strength and the
proportioning   required additional rate ($\alpha \cdot q_{udd}$).

Conversion from the conventional unit system to the SI unit system in this manual treats the gravity acceleration as 10 m/s^2.

CHAPTER 1

Outline of the Premixing Method

1.1 INTRODUCTION

In Japan it has been common practice to execute ground improvement of reclaimed soft-soil ground after reclamation of the ground has been completed. When land use plans and execution plans for the reclaimed land are known, it is possible to shorten the overall execution period and lower execution costs by planning and executing the reclamation work in anticipation of the ground improvement measures forecast to be performed after the completion of the reclamation. In the case of ground where liquefaction might occur, measures to prevent liquefaction are usually taken, but if materials that will not liquefy are used for the reclamation, post-reclamation ground improvement is unnecessary.

In recent years, it has become increasingly difficult to obtain good quality landfill materials that will not liquefy. One approach to overcoming this problem is to perform the reclamation work with a landfill material that has been treated by a method that prevents it from liquefying.

Considering the cost and time required to perform post-reclamation ground improvement work, in many cases it is beneficial to take counter-measures beforehand. The premixing method described herein has been developed in line with this new approach. It is performed by adding small quantities of a stabilizer (cement etc.) and a separation inhibitor to the soil in advance to make treated soil. This treated soil is transported to, and placed at, the stipulated location to form stable ground. When executed below the surface of the water, density control performed by compaction etc. is omitted and the addition of the cement increases the stability of the ground.

This method was originally developed as a way to prevent the liquefaction of sandy ground, but it is also used to lower the earth pressure behind quaywalls and similar structures. Improving soil before placing it is a well established approach. A typical example is the use of soil cement during road paving work. This is the application of the method as only one part of a larger execution. Its use to improve the entire ground formed by a reclamation project was a new challenge intended to resolve design problems such as guaranteeing ground stability. New problems that had to be overcome to execute this method were the development of methods of adding and mixing the stabilizer etc., and methods of placing the treated soil underwater. This method has been named the "premixing method" to distinguish it from existing construction methods because these problems were resolved by the premixing method. When it is categorized based on ground improvement principles, it is placed in the same category as the deep mixing method, which is also a chemical hardening method. Since the basic research began, the development of the method has taken 10 years. The Design of Ground Improvement by the Premixing Method (Coastal Development Institute of Technology) was published in 1989. This technical material is published based on experience gained executing the method.

1.2 DEVELOPMENT OF THE PREMIXING METHOD

1. *Soil treated*
1.1 The soil to be treated is limited to sandy soil with a sand content larger than 80%.
1.2 Because the properties of volcanic ash soil differ from sandy soil, it is treated by this method after its post-treatment engineering properties have been clarified.
2. *Characteristics of the method*
2.1 *Advantages*
 – It shortens the construction period because post-reclamation ground improvement is unnecessary.
 – The strength of the treated ground can be set at any value within a certain range.
 – Large-scale execution or deep water execution can be conducted based on various reclamation methods including chute reclamation and direct reclamation performed by bottom dump barges.
 – The soil and the stabilizer used for the reclamation can be mixed continuously in large quantities using a belt conveyor.
 – It can be performed using existing facilities, vessels, etc.
 – It is quieter and produces less vibration than vibration and compaction methods.
 – Because compaction is unnecessary, it has little effect on quaywalls, etc.
 – Construction by products such as dredged soil can be utilized for the soil to be treated.
2.2 *Disadvantages*
 – Variation in improvement effectiveness and the scattering of the strengthening achieved occur according to the type of soil used for the reclamation.
 – Some cost is associated with the stabilizer.
 – Its influence on the water quality and other environmental impacts must be considered.

– It can only be applied to reclamation and to soil that is to be excavated and replaced.

3. *Uses and applications of the method.* The premixing method is used on newly reclaimed ground to prevent liquefaction and reduce earth pressure. As shown in Figure 1.1, in this manual, reclamation ground refers to backfill behind quaywalls and bulkheads, cell and caisson fill, and the replacement and backfilling executed following seabed excavation. In addition to the applications shown in Figure 1.1, this method can also be applied to achieve ground strength needed by structures to be placed on reclamation ground. In this case, the strength of the treated ground must be accurately evaluated.

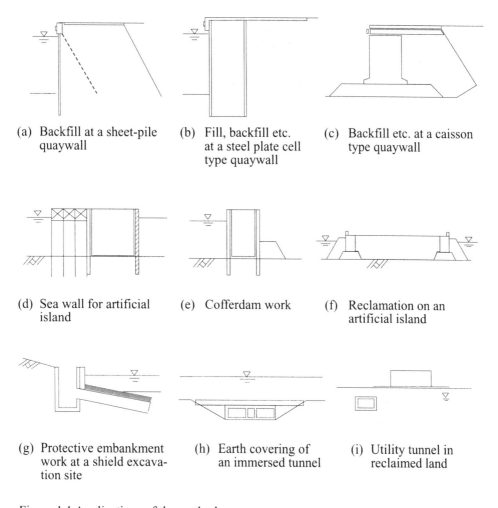

(a) Backfill at a sheet-pile quaywall

(b) Fill, backfill etc. at a steel plate cell type quaywall

(c) Backfill etc. at a caisson type quaywall

(d) Sea wall for artificial island

(e) Cofferdam work

(f) Reclamation on an artificial island

(g) Protective embankment work at a shield excavation site

(h) Earth covering of an immersed tunnel

(i) Utility tunnel in reclaimed land

Figure 1.1 Applications of the method.

1.3 STABILIZATION MECHANISM AND HARDENING AGENTS

1. *Stabilization method.* Table 1.1 shows the methods used according to conditions at the time of the mixing and placing of the soil.
1.1 *Dry method.* Adding the cement or other stabilizer and the separation inhibitor to sandy material at or below the natural water content without adding any additional water.
1.2 *Wet method.* Using dredged soil, excavated soil or other soil with a relatively high water content without modification. Other than differences between their mixing methods, this method and the dry method are identical.
1.3 *Slurry method.* Adding water along with stabilizer, high polymer material, and a fine fraction to obtain flowability and resistance to separation.

Table 1.1 Treatment methods of premixing method.

Methods	Untreated soil	Mixing Method	Treated Soil Transportation, Unloading, and Placing Method
Dry method	Sandy soil, dry to wet: at or below natural water content (low water content)	Mixed by belt conveyor connection systems	- Transported on dump trucks then spread on the ground by a bulldozer etc. - Transported by a belt conveyor then placed with a double-wall chute (mixed on the ocean surface) - Transported by soil barge and dumped directly into the ocean.
Wet method	Sandy soil, wet (medium water content)	Mechanically mixed in a 2-shaft pugmill mixer	Transported on dump trucks then spread on the sea bottom by bucket
Slurry method	Sandy soil containing clay and a fine fraction	Mechanically mixed in a 2-shaft pugmill mixer	Transported by a conveying pump then placed with a tremie pipe

* The slurry method is not described in technical material.

2. *Improvement principles.* In the premixing method, a stabilizer like cement is employed the cohesion of the soil used for the reclamation work is increased based on chemical hardening of the soil and the stabilizer. The increase in cohesion is dependent on the hydration reaction of the stabilizer and the chemical reactions that continue for a long time afterward.

Since the soil that has been stabilized by the addition of the stabilizer is placed directly underwater, one concern is the separation of the stabilizer in the water. This is prevented by the method described below.

First the stabilizer is added to and mixed with the soil so that it adheres to the surface of the soil particles. Then the separation inhibitor is added to the mixture. Adding the separation inhibitor to the mixture afterward, means that it will coat the mixed bodies that consist of soil particles and stabilizer to prevent their separation in the water.

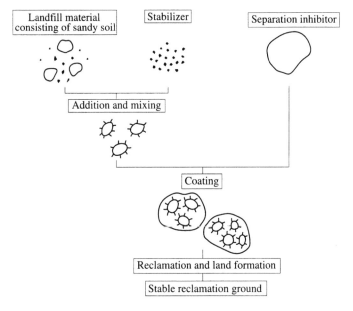

Figure 1.2 Stabilizer and separation inhibitor addition and mixing mechanism. (Coastal Development Institute of Technology, 1989)

Photograph 1.1 shows the state of the sand particles before treatment and Photograph 1.2 shows the adhesion of the cement on the soil particles after treatment. This illustrates how the cement has formed ettringite and hardened through the pozzolanic reaction around the points of contact of the soil particles.

Photograph 1.1 Microphotograph of sand particles before treatment (Sengenyama pit sand).

Photograph 1.2 Microphotograph of sand particles after treatment (Sengenyama pit sand).

3. *Properties of the treated soil and the treated ground.* (Coastal Development Institute of Technology, 1989) The following are the basic benefits of treated soil and treated ground prepared by adding a small amount of stabilizer to sandy soil.

The following properties, however, are based on past test data, and there is a possibility that exceptional properties are encountered according to the type of soil used.

3.1 *Unconfined compression properties*
   – The unconfined compressive strength varies according to the type of stabilizer added and increases as the percentage added and the dry density rise.
   – The unconfined compressive strength varies according to the curing temperature and increases with age.
   – The increase in strength varies according to the type of soil used as the parent material.
   – Failure strain falls within a range from 0.5% to 2% regardless of the type of parent material.

3.2 *Triaxial compression properties*
   – The stress-strain relationship tends to be similar regardless of the type of soil used.
   – The principle of effective stress is applicable and failure criteria is established regardless of drainage conditions.
   – The angle of shear resistance of the treated soil is almost equal to that of the untreated soil, but the cohesion of the treated soil is higher. Therefore, the treated soil has the properties of a $c - \phi$ material.
   – The degree of brittleness of the treated soil varies according to the type of soil used as the parent material.
   – The unconfined compressive strength of the treated soil is almost equal to the unconfined drained test strength.

– If the angle of shear resistance (ϕ_d) of the untreated soil and the unconfined compressive strength (q_{ud}) of the treated soil are obtained by testing, it is possible to estimate the shear strength of the treated soil.

3.3 *Cyclic strength properties.* The cyclic strength is increased by the addition of the stabilizer, and the relationship of the unconfined compressive strength with the cyclic shear stress ratio τ_d/σ_c' is an almost linear correlation.

3.4 *Dynamic deformation properties.* The shear modulus after treatment (value under a strain amplitude of 10^{-6}) varies according to the percentage of stabilizer added, and it becomes approximately 2 to 6 times that before treatment. No particular difference between the damping ratio before and after treatment has been observed.

3.5 *Elastic wave velocity.* There is a correlation between the shear wave velocity and the unconfined compressive strength.

3.6 *Coefficient of permeability.* Its coefficient of permeability is about one order lower than that of untreated soil.

References

Coastal Development Institute of Technology, 1989. *Design of treated ground using the premixing method*: 1-2 (in Japanese).

CHAPTER 2

Factors Influencing the Increase in Strength

2.1 INTRODUCTION

The strength of the treated soil prepared by the premixing method is influenced by a variety of factors, as follows:
   (a) Type of stabilizer and percentage of stabilizer added
   (b) Curing temperature and age
   (c) Type of parent material treated
This chapter presents the relationship between the principal factors and the unconfined compressive strength (Table 2.1). These relationships are also related to mixing conditions and execution method.

Table 2.1 Factors influencing the increase in strength.

[1] Type and percentage added of stabilizer	Figure 2.1
[2] Curing temperature and age	Figure 2.2, Figure 2.3
[3] Type of parent material treated	Figure 2.4, Figure 2.5

2.2 TYPE OF STABILIZER AND PERCENTAGE OF STABILIZER ADDED

Figure 2.1 compares the strengths of Rokko masa soil treated with ordinary portland cement and with slag cement. It reveals that slag cement results in greater strength. And as the percentage of stabilizer added is increased, the curves that represent the relationship of the unconfined compressive strength with the dry density rise almost in parallel.

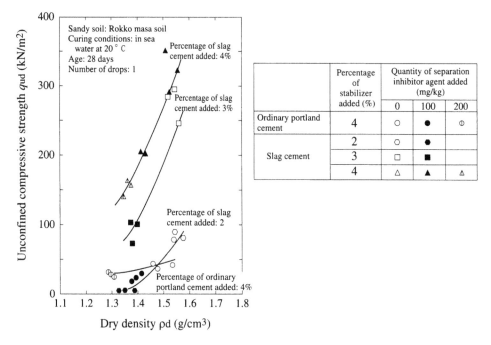

Figure 2.1 Comparison of unconfined compressive strength when type of stabilizer and percentage of the stabilizer added are varied (Zen et al., 1987).

2.3 CURING TEMPERATURE AND AGE

Figure 2.2 shows the improvement in unconfined compressive strength of a soil prepared by improving Rokko masa soil (maximum grain size 4.76 mm) with slag cement and varying the percentage of stabilizer added (5%, 10%), the curing temperature (10 °C, 20 °C, and 30 °C) and the age (7 days, 28 days, 90 days). According to Figure 2.2, the unconfined compressive strength of treated soil increased as the curing temperature rose and also increased with age. Figure 2.3 shows the relationship of the ratio of the strength to that at 28 days with age and temperature. It is found, from Figure 2.3, that the strength ratio was 0.35 at 7 days and 1.20 at 90 days. It was almost stable at 90 days.

These findings indicate that the criterion for age should be 28 days for execution control. And when the curing temperature and the age are varied, it is possible to estimate their influence on the strength by determining the type of stabilizer to be used and obtaining the strength at 7 days for a standard curing temperature of 20 °C.

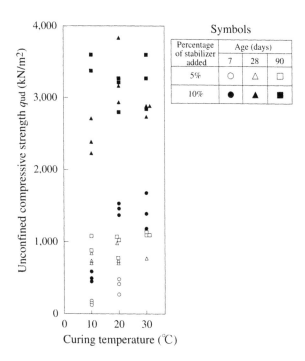

Figure 2.2 Curing temperature and unconfined compressive strength (Mori et al.).

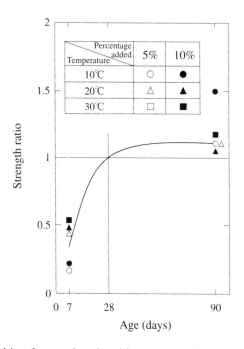

Figure 2.3 Relationship of strength ratio with age and curing temperature (Mori et al.).

2.4 TYPE OF SOIL TREATED

Figure 2.4 shows the relationship of the unconfined compressive strength with the dry density for a case where masa soil from different locations with the grain size distributions shown in Figure 2.5 are improved by the addition of 3% of slag cement. Other studies for various grain size distributions of landfill soils, such as different gravel fraction contents and adjusted fines content, were also performed. Through these studies, the following relationships between the type of soil and the unconfined compressive strength were confirmed.

1. The increase in strength of masa soil made of weathered granite varies if it is obtained from different locations.

2. If the gravel fraction content varies, the strength increase also varies.

3. The strength increase varies according to the properties of the fines content and the percentage of cement added.

Therefore, when designing the improvement of ground using this method, it is essential to perform soil properties testing and mix proportion testing for each of the landfill soils to be used .

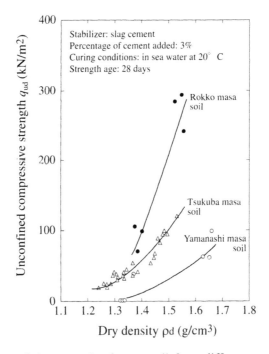

Figure 2.4 Comparison of the strength of masa soil from different regions (Zen et al., 1987).

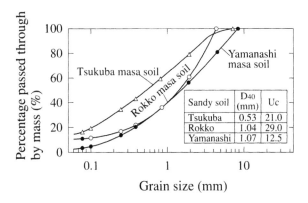

Figure 2.5 Grain size distribution (Zen et al., 1987).

References

Mori K. et al. Development of the premixing method (Part 1), basic laboratory experiments. *JDC Corporation Technical Research Report* (in Japanese).

Zen K. et al., 1987. Study on a reclamation method with cement-mixed sandy soils – fundamental characteristics of treated soils and model tests on the mixing and reclamation. *Technical Note of the Port and Harbour Research Institute*, No. 579 (in Japanese).

CHAPTER 3

Engineering Properties of Treated Soil

3.1 INTRODUCTION

This chapter is a detailed explanation of the engineering properties of treated soil that are unconfined compression properties, triaxial compression properties, cyclic strength, dynamic deformation properties, elastic wave velocity, and coefficient of permeability. In addition, this chapter explains model experiments and field observations performed to study the liquefaction properties and earth pressure reduction.

The following is a summary of the basic properties of lean treated soil and ground, where sandy soil is used as the parent material. But the following properties are based on past test data, and there is a possibility that exceptional properties are encountered according to the type of soil used.

Ұ Unconfined compression properties
1. The unconfined compressive strength varies according to the type of stabilizer added and increases as the percentage added and the dry density rise.
2. The unconfined compressive strength varies according to the curing temperature and increases with the age.
3. The increase in the strength varies according to the type of parent material.
4. The failure strain falls within a range from 0.5% to 2% regardless of the type of parent material.

Ұ Triaxial compression properties
1. The stress – strain relationship tends to be similar regardless of the parent material.
2. The principle of effective stress is applicable and failure criteria is established regardless of drainage conditions.
3. The angle of shear resistance of the treated soil is almost equal to that of the untreated soil, but the cohesion of the treated soil is higher. Therefore, the treated soil has the properties of a c -ϕ material.
4. The brittleness of the treated soil varies according to the type of parent material.
5. The unconfined compressive strength of the treated soil is almost equal to the

unconfined drained test strength.

6. If the angle of shear resistance (ϕ_d) of the untreated soil and the unconfined compressive strength (q_{ud}) of the treated soil are obtained by test, it is possible to estimate the shear strength of the treated soil.

¥ Cyclic strength properties

The cyclic strength is increased by the addition of the stabilizer, and the relationship of the unconfined compressive strength with the cyclic shear stress ratio, τ_d / σ_c' indicates an almost linear.

¥ Dynamic deformation properties

The shear modulus after treatment (value at a strain amplitude of 10^{-6}) varies according to the percentage of stabilizer added and becomes approximately 2 to 6 times that before treatment. No particular difference of the damping ratio before and after treatment has been observed.

¥ Elastic wave velocity

There is a correlation between the shear wave velocity and the unconfined compressive strength.

¥ Coefficient of permeability

The coefficient of permeability of the treated soil is about one order lower than that of the untreated soil.

3.2 MECHANICAL PROPERTIES

1. *Strength properties*

1.1 *Unconfined compressive properties.* Factors influencing the unconfined compressive strength of treated soil include the type of stabilizer used, the percentage of stabilizer added, the curing temperature, the age, and the type of landfill soil used. These are as explained in the previous chapter. In this chapter, the failure strain and the modulus of deformation are explained as the unconfined compression properties of treated soil.

1.1.1 *Failure strain and modulus of deformation.* Figure 3.1 shows the relationship between the failure strain and the unconfined compressive strength of treated soil. It presents the results for six kinds of landfill soil, revealing that the failure strain ranges from 0.5% to 2.0% regardless of the type of soil. The relationship between the unconfined compressive strength and the modulus of deformation E_{50} is proportional, as shown in Figure 3.2.

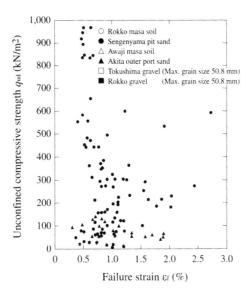

Figure 3.1 Failure strain – unconfined compressive strength relationship of treated soil.

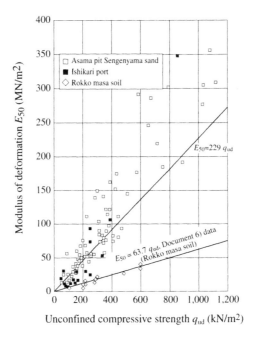

Figure 3.2 Relationship of unconfined compressive strength with modulus of deformation of treated soil.

1.2 *Triaxial compression properties.* The following are the principal characteristics of the triaxial compression (CD, $\overline{\text{CU}}$) properties of untreated and treated (percentage of stabilizer added: 5.5%) Rokko masa soil and Akita outer port sand.

1.2.1 The deviator stress under a constant strain increases as the effective confining pressure rises and the density rises.

1.2.2 Regarding the deviator stress and the effective stress path on the mean effective principal stress plane, the point that indicates the peak strength is on a straight line regardless of whether it is drained or undrained, and the residual strength is similarly on a straight line. If represented in terms of the effective stress, the failure criteria are uniquely established regardless of whether it is drained or undrained (Fig. 3.3). The same tendencies have been obtained from research by Maeda et al., 1988.

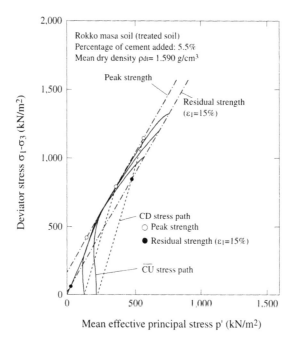

Figure 3.3 Effective stress path of treated soil (Zen et al., 1988).

1.2.3 The rupture envelope of Mohr of treated soil moves upward in parallel to the envelope of untreated soil in both the CD and $\overline{CU}$ tests (Figs 3.4 and 3.5).

1.2.4 The relationship of the strength parameters (cohesion, angle of shear resistance) with the relative density of treated soil at peak strength is characterized as follows. The cohesion increases as the relative density increases, while the angle of shear resistance is either a little larger (in the case of Rokko masa soil) than or equal (in the case of Akita outer port soil) to that of untreated soil (Fig. 3.6).

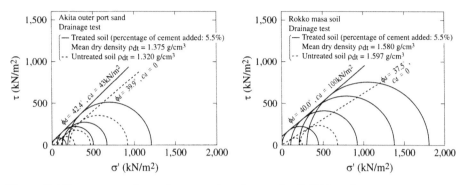

Figure 3.4 Rupture envelope of untreated and treated soil (CD test).

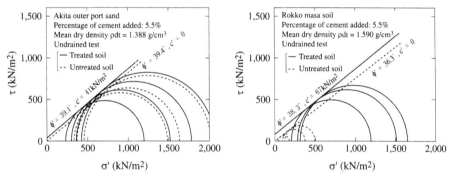

Figure 3.5 Rupture envelope of untreated and treated soil ($\overline{CU}$ test).

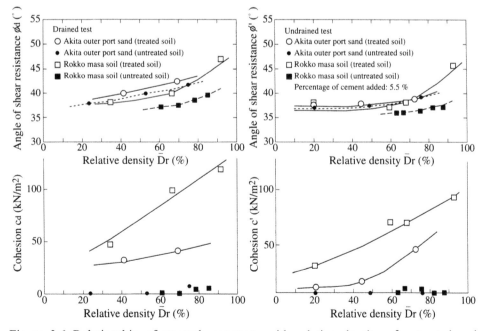

Figure 3.6 Relationship of strength constants with relative density of untreated and treated soil of peak strength.

1.2.5 The cohesion of treated soil during residual strength is lower than that of peak strength, but the angle of shear resistance remains almost constant without falling (Figs 3.7 and 3.8).

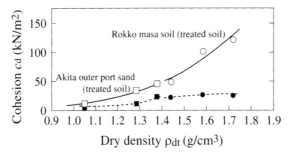

Figure 3.7 Cohesion of treated soil at peak strength in a triaxial CD test (Maeda et al., 1988).

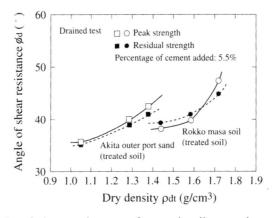

Figure 3.8 Angle of shear resistance of treated soil at peak strength and at residual strength in a triaxial CD test (Maeda et al., 1988).

1.2.6 The strength and deformation properties of the treated soil obtained by unconfined compression test tend to resemble that from drained triaxial test of $\sigma_c'=0$kN/m^2 (Fig. 3.9), and unconfined compressive strength approximates the drained strength under unconfined condition (Fig. 3.10).

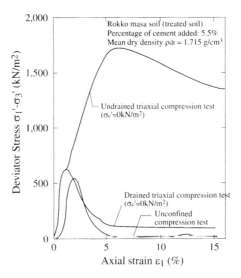

Figure 3.9 Evaluation of unconfined compressive strength (Maeda et al., 1988).

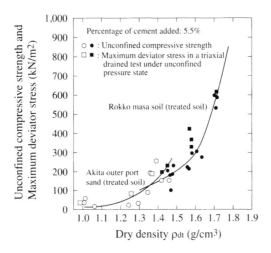

Figure 3.10 Relationship of the unconfined compressive strength with the maximum deviator stress in drained triaxial compression test under unconfined pressure (Maeda et al., 1988).

2. *Dynamic properties*

2.1 *Cyclic strength properties.* Figure 3.11 presents an example of the relationship of the cyclic shear stress ratio (τ_d / σ_c') with the number of cycles (N_f) based on cyclic triaxial tests of treated (ordinary portland cement; percentage added from 2.6 to 2.9%) and untreated Awaji masa soil and Akita outer port sand. Figure 3.11 reveals that the addition of a small quantity of stabilizer increases the cyclic shear stress ratio and remarkably increases the cyclic shear strength.

Figure 3.12 plots the relationship of the cyclic strength with the unconfined compressive strength q_{ud}, showing that although there is scattering, they are almost in a linear correlation. The cyclic strength where treated soil has been prevented from liquefying is estimated from the unconfined compressive strength using this relationship. Yamamoto et al. set the curing period at 1 day, while the results of other tests set it at 28 days, but both share common tendencies. These tests reveal that the liquefaction strength is closely related to the effective grain size D_{10}.

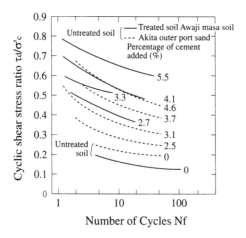

Figure 3.11 Cyclic strength of treated soil (Awaji Masa Soil, Akita Outer Port Sand) (Zen et al., 1987).

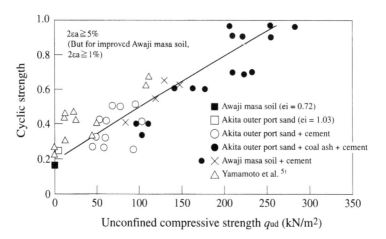

Figure 3.12 Correlation of cyclic strength with unconfined compressive strength.

2.2 *Element test.* Figure 3.13 shows the results of cyclic triaxial tests on untreated soil ($\tau_c / \sigma_c' = 0.12$) and treated soil ($\tau_c / \sigma_c' = 0.80$). In the untreated soil case, the excess pore water pressure ratio reached 1.0 and the specimen liquefied; deformation occurred both in a compressive and an extensive direction. In the treated soil case, however, deformation occurred only in the extensive direction and failure occurred while the excess pore water pressure ratio was less than 0.1. In this way, for a stabilized specimen in a cyclic triaxial test, if the percentage of stabilizer added is large enough, liquefaction does not occur but tensile failure like concrete occurs.

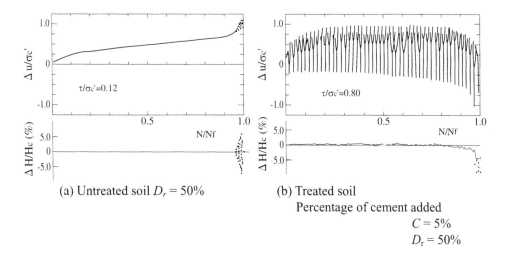

(a) Untreated soil $D_r = 50\%$ (b) Treated soil
Percentage of cement added
$C = 5\%$
$D_r = 50\%$

Figure 3.13 Time histories of the excess pore water pressure ratio ($\Delta u/\sigma_c'$) and the axial strain ($\Delta H/H_c$).

Figure 3.14 shows the relationship of the cyclic strength with the unconfined compressive strength of treated soil. In this figure, specimens resulting in tensile failure are distinguished from those resulting in liquefaction failure. It shows that if the unconfined compressive strength exceeds 50 to 100 kN/m², the specimen tends result in tensile failure rather than liquefaction failure. This fact reveals that the strength criteria for the treated soil that has been prevented from liquefying is $q_{ud} = 50$ to 100 kN/ m² or more.

Photograph 3.1 Cyclic triaxial test for a block sample (tensile failure occurs).

Photograph 3.2 Specimen after test (no indications of liquefaction).

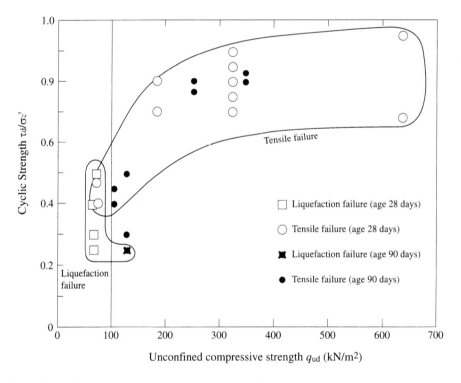

Figure 3.14 The relationship between unconfined compressive strength and cyclic strength.

2.3 *Dynamic deformation properties*

2.3.1 *Shear modulus.* Figure 3.15 is an example of the measurement of the shear modulus G of improved Awaji masa soil (ordinary portland cement, percentage added: 5.5%). For comparison, the shear modulus of untreated soil is also shown. Figure 3.16 presents the results of a comparison of the shear modulus G_0 (value at a strain amplitude of 10^{-6}) before and after treatment. The increase of the shear modulus of the treated soil varies according to the percentage of stabilizer added, but it generally ranges between 2 and 6 times. Figure 3.17 reveals the almost linear correlation of the shear modulus G_0 with the unconfined compressive strength.

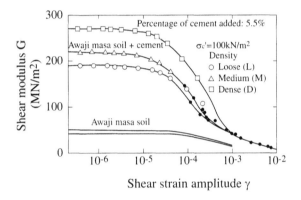

Figure 3.15 The relationship between share modulus and strain amplitude of treated soil (Awaji Masa Soil, percentage of stabilizer added: 5.5%) (Zen et al., 1987).

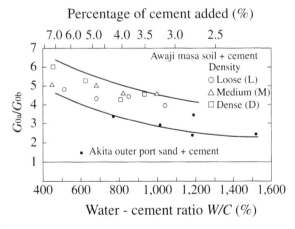

Figure 3.16 Change of the share modulus before and after treatment (Zen et al., 1987).

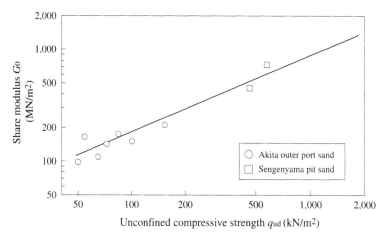

Figure 3.17 Correlation of unconfined compressive strength with shear modulus (Zen et al., 1987).

2.3.2 *Damping ratio.* Figure 3.18 shows the damping ratio measured from the same specimen as in Figure 3.15. Figure 3.19 compares the damping ratio before and after treatment, and reveals no evident difference.

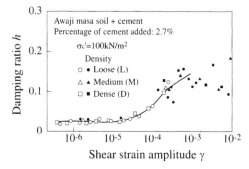

Figure 3.18 The relationship between damping ratio and strain amplitude of treated soil (Awaji masa soil, percentage of cement added: 2.7%) (Zen et al., 1987).

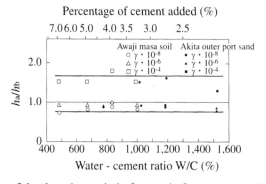

Figure 3.19 Change of the damping ratio before and after treatment (Zen et al., 1987).

2.4 *Shear wave velocity*. Figure 3.20 presents the relationship between unconfined compressive strength and shear wave velocity of treated soil when between 3.0% and 7.0% stabilizer was added to sandy soil and three kinds of gravelly soil with a maximum grain size of 50.8 mm. Figure 3.20 shows that there is a unique relationship between the unconfined compressive strength and the shear wave velocity, demonstrating that it is possible to predict the unconfined compressive strength from the shear wave velocity.

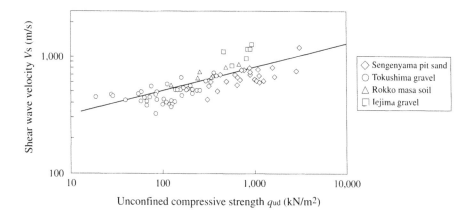

Figure 3.20 Unconfined compressive strength – shear wave velocity relationship.

3. *Consolidation properties*

3.1 *Coefficient of permeability and drainage properties (coefficient of consolidation)*. Table 3.1 compares the coefficients of permeability of untreated and treated Tokushima gravel, Rokko masa soil, and Tsukuba masa soil. The values for the treated soils were obtained by in-situ permeability tests on reclamation ground. Table 3.1 shows that the coefficient of permeability of treated soil is approximately one order lower than that of untreated soil. But the coefficient of permeability is not low enough to permit the use of treated soil as a cut-off wall. The drainage properties of the ground under loading are, as well as consolidation, influenced not only by the effects of the coefficient of permeability but also by the effects of the coefficient of volume compressibility, and are governed by a coefficient of consolidation that is represented as a ratio of the coefficient of permeability to the coefficient of volume compressibility. As shown in Figure 3.14, the shear modulus of soil treated with percentage of cement added of 5% rises to about 5 times that of untreated soil. Consequently, the coefficient of permeability of treated soil is about 1 order lower, but the coefficient of volume compressibility is found to decline by about 1/5, and the coefficient of consolidation decreases by about 1/2. These facts show that the drainage properties of treated soil under loading are lower than those of untreated soil but not by as much as an order lower.

Table 3.1 Comparison of the coefficients of permeability of untreated soil and treated soil.

Type of Soil Percentage of Stabilizer Added	Test Measurement Method		Untreated Soil k (cm/s)	Treated Soil k (cm/s)
Tokushima gravel $C = 4\%$	Laboratory	Loose density	$2\sim3\times10^{-2}$	—
		$Ec\times1/2$	8×10^{-3}	—
		$Ec\times1$	4×10^{-3}	—
	In-situ (on reclamation ground)	$0\sim0.5$m	1×10^{-1}	2×10^{-2}
		$0\sim1.5$m	1×10^{-1}	8×10^{-3}
Rokko masa soil $C = 5.5\%$	Laboratory	Loose density	5×10^{-3}	—
	In-situ (on reclamation ground)	$0.6\sim1.7$m	—	$3\sim4\times10^{-4}$
		$0.6\sim3.2$m	—	2×10^{-4}
Tsukuba masa soil $C = 5.5\%$	Laboratory	Loose density	3×10^{-3}	—
	In-situ (on reclamation ground)	$0.6\sim1.5$m	—	3×10^{-4}
		$0.6\sim2.3$m	—	2×10^{-4}
		$0.6\sim2.8$m	—	2×10^{-4}

Ec: Compactive effort during a Proctor compaction test (m·kN/m^3)

3.3 EFFECTS OF LIQUEFACTION MEASURES

1. *Model test.* Two shaking table tests were executed on grounds of Niigata eastern port sand, one untreated and the other treated with 1% ordinary Portland cement, according to the schematic diagram presented in Figure 3.21. The test was performed under stage loading of 20 sinusoidal waves, excited at acceleration of 50gal, 100gal, 150gal, 200gal, and 250gal. Figure 3.22 shows the distribution of the maximum excess pore water pressure, and Figure 3.23 shows the time history of excess pore water pressure of the untreated and treated ground excited at an acceleration of 100gal. The results of the test reveal that in the case of the ground treated with 1% cement stabilizer, almost no excess pore water pressure is generated, far different from the untreated soil case. They also confirm that, even if excited at an acceleration of 250gal, the treated ground does not liquefy.

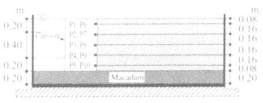

Figure 3.21 Schematic diagram of the test.

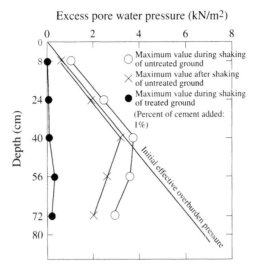

Figure 3.22 Distribution of maximum excess pore water pressure.

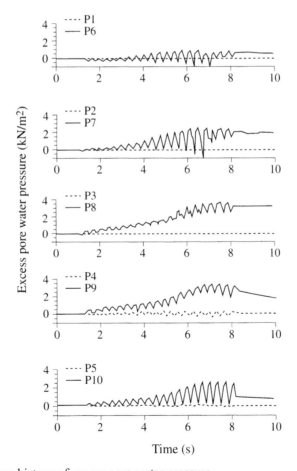

Figure 3.23 Time history of excess pore water pressure.

3.4 EARTH PRESSURE REDUCTION EFFECTS

1. *Model test.* To study the earth pressure reduction effects obtained by the premixing method, the earth pressure at rest and the active earth pressure in ordinary condition were studied by performing a static loading test using the model soil tank shown in Figure 3.24. The percentage of cement added to make the treated ground ranged between 1%, 3%, and 5%. Untreated ground was also tested, and the results for each case were compared. The earth pressure at rest was measured twice, once at the time the ground was prepared and again after it had been cured for 7 days. The active earth pressure was measured by loading a surcharge of 100 kN/m^2 after curing and moving a movable wall.

Figure 3.25 shows the results of the measurements of the earth pressure at rest when the ground was prepared and after 7 days of curing. The theoretical values in Figure 3.25 were obtained using the soil constants in Table 3.2 based on Jàky's coefficient of earth pressure at rest, $K_0 = 1\text{-}sin\phi$. The values γ_t, q_{ud}, ϕ_d, and c_d in Table 3.2 for each percentage of cement added were obtained from mix proportion tests performed separately, and $K_0 (=\sigma_h'/\sigma_v')$ is a value measured from the load increment during surcharge loading after the completion of curing. In addition, K_a is the coefficient of active earth pressure obtained from Rankine's active earth pressure equation.

Figure 3.25 demonstrates that the earth pressure at rest after the completion of curing is smaller than it was when the ground was prepared, confirming the effects of the increase in cohesion by cement. Table 3.2 shows that the coefficient of earth pressure at rest $K_0 (=\sigma_h'/\sigma_v')$ obtained from the increase in earth pressure after curing falls as the percentage of cement added increases, confirming its earth pressure reduction effects. Figure 3.26 shows the measured values of the active earth pressure, while the theoretical values were obtained from the soil constants in Table 3.2 and from the Rankine - Resal active earth pressure equation. Up to a percentage of cement added of 3%, the earth pressure reduction effects of adding cement are not clearly evident, but at a percentage of cement added of 5%, the active earth pressure falls substantially.

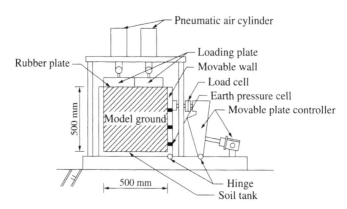

Figure 3.24 Schematic diagram of the test.

Table 3.2 Soil constants.

Soil Constants	Untreated Soil	Treated Soil $C = 1\%$	Treated Soil $C = 3\%$	Treated Soil $C = 5\%$
		Test Cases		
γ_t (kN/m^3)	18.8	18.4	18.5	18.4
q_u (kN/m^2)	—	13	28	94
ϕ_d (°)	38.5	36.4	36.5	36.8
c_d (kN/m^2)	0	3.5	7.5	17.9
K_0 (=1 – $\sin \phi_d$)	0.377	0.407	0.405	0.401
K_0 (Measured)	0.30	0.27	0.21	0.06
K_a (calculated from ϕ_d)	0.233	0.255	0.254	0.251

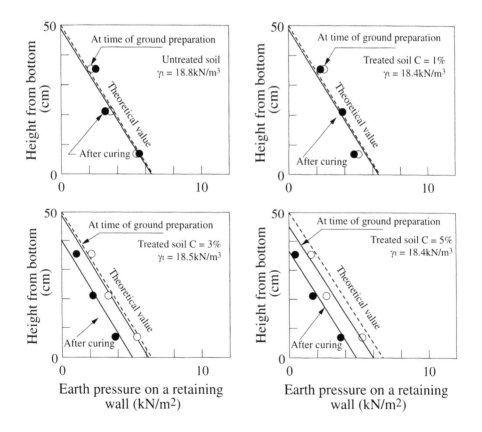

Figure 3.25 Earth pressure at rest.

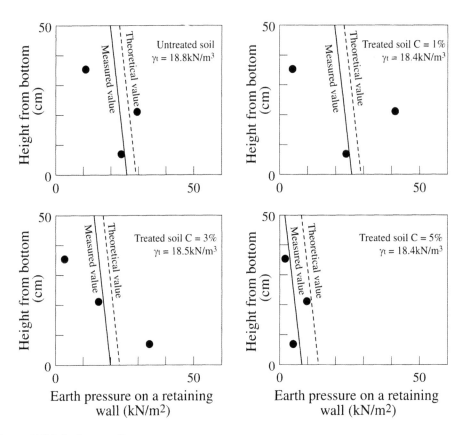

Figure 3.26 Active earth pressure.

2. *Field observations.* Figure 3.27 shows the results of measurements of earth pressure against a wall during execution on Rokko Island. In this execution, the premixing method was adopted for reclamation from G.L. – 12 m to G.L. + 1.7 m, using 7.5% cement added masa soil as the landfill material. Figure 3.27 shows that the coefficient of earth pressure ranged from 0.1 to 0.3, confirming that the method reduced the earth pressure.

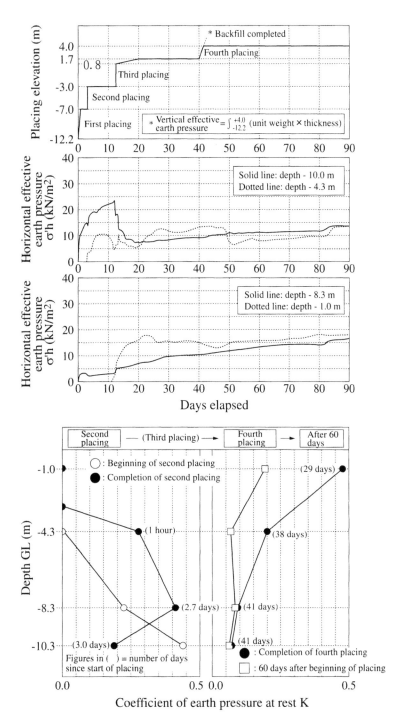

Figure 3.27 Results of measurement of earth pressure against wall during execution on Rokko Island.

References

Coastal Development Institute of Technology, 1989. *Design of treated ground using the premixing method*: 1-51 (in Japanese).

Oikawa K. et al., 1997. Site investigation on the earth pressure distribution in the cement treated ground. *Proceedings of the 32nd Japan National Conference on Geotechnical Engineering*, 1: 303-304 (in Japanese).

Yamamoto T. et al., 1996. Effects of grain size characters on the effectiveness of cement fixing agent method for preventing liquefaction of sand to silt deposits. *Journal of Geotechnical Engineering*, No. 541/III-35: 133-146, (in Japanese).

Zen K. et al., 1987. Study on a Reclamation Method with Cement-mixed Sandy Soils – Fundamental Characteristics of Treated Soils and Model Tests on the Mixing and Reclamation – *Technical Note of the Port and Harbour Research Institute* Ministry of Transport, Japan, No. 579 (in Japanese).

Zen K. et al., 1990. Strength and Deformation Characteristics of Cement Treated Sands Used for Premixing Method. *Report of the Port and Harbour Research Institute* Ministry of Transport, 29, No. 2 (in Japanese).

Zen K. et al., 1991. Shaking table test on liquefaction resistance of premixed soil. *Proceedings of the 46th Annual Conference of the Japan Society of Civil Engineers*, 3: 216-217 (in Japanese).

Zen K. et al., 1992. Model experiment concerning earth pressure at rest of soil treated by the premixing method. *Proceedings of the 47th Annual Conference of the Japan Society of Civil Engineers*, 3: 972-973 (in Japanese).

Zen K. et al., 1999. *"8.6.4 the premixing method", Handbook of Geotechnical Engineering*: 1242-1245. The Japanese Geotechnical Society (in Japanese).

CHAPTER 4

Execution Examples

4.1 EXECUTION TABLE

Up to 1998, the premixing method was used 5 times and the quantity of treated soil by this method reached 1,264,000 m^3. Table 4.1 depicts these executions and Figure 4.1 shows their locations. This document is a summary of outlines of the work performed during these 5 improvements.

Table 4.1 Premixing method execution table.

Section Number	Project	Purpose of Improvement	Design Strength	Quantity Treated
4.2	Artificial Island on the Kisarazu Side of the Tokyo Aqualine	- Prevention of liquefaction - Guarantee of bearing capacity	Unconfined compressive strength $q_{ud} = 400$ kN/m^2 (age: 91 days)	8000 m^3 (preliminary) 425,000 m^3 (preliminary)
4.3	Niigata Airport	- Prevention of liquefaction	Unconfined compressive strength $q_{ud} = 100$ kN/m^2 (age: 28 days)	200 m^3 (preliminary) 118,510 m^3 (preliminary)
4.4	New East Harbor in the Port of Ishikari – 7.5 m quaywall	- Prevention of liquefaction	Unconfined compressive strength $q_{ud} = 50$ kN/m^2 (age: 28 days)	1,160 m^3 (preliminary) 57,000 m^3 (preliminary)
4.5	Port of Kobe, Rokko Island – 10 m quaywall etc.	- Earth pressure reduction	Unconfined compressive strength $q_{ud} = 100$ kN/m^2 C = 50 kN/m^2, ϕ=30°	620,000 m^3
4.6	Port of Kobe, Rokko Island, Ferry Quaywall (RF-3)	- Earth pressure reduction	Unconfined compressive strength $q_{ud} = 100$ kN/m^2 (Age 28 days) C = 50 kN/m^2, ϕ=30°	34,300 m^3
Total treated soil				1,264,170 m^3

Figure 4.1 Execution locations.

4.2 ARTIFICIAL ISLAND ON THE KISARAZU SIDE OF THE TOKYO BAY AQUALINE

1. *Outline of the work*. The artificial island on the Kisarazu side of the Tokyo Bay Aqualine is, as shown in Figure 4.2, the connection point between the shield tunnel on the Kawasaki side and the bridge on the Kisarazu side.

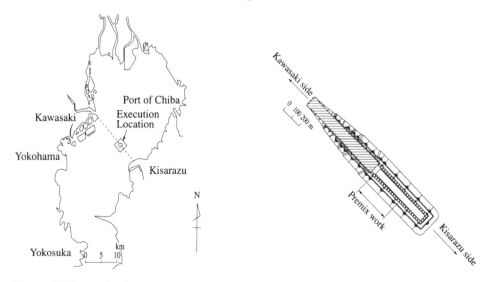

Figure 4.2 Execution location.

The artificial island was constructed by enclosing a 400 m×60 m water area with a depth of 25 m with a steel sheet pile cell bulkhead (Fig. 4.3), and then reclaiming the land inside the bulkhead with the premixing method. As shown in Figure 4.4, the reclamation ground was dried out and a box culvert and L-shaped retaining wall were build on it. The design strength of the reclamation ground was the unconfined compressive strength of 400 kN/m^2 (age 28 days).

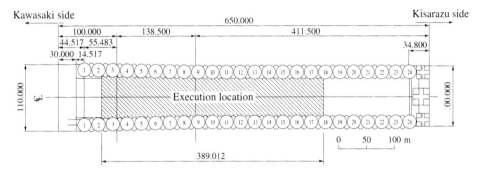

Figure 4.3 Plane diagram of the execution.

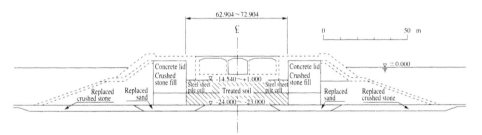

Figure 4.4 Section diagram of the execution.

2. *Quantity of work*

2.1 *Execution quantities*

- Preliminary field test: 8,000 m^3
- Execution: 425,000 m^3

2.2 *Mix proportion at job site.* The following is the mix proportion per 1 m^3 with the required additional rate of 1.0 when the design q_{ud} = 400 kN/m^2 (age: 28 days).

Soil: Sengenyama pit sand   1,330 kg (dry weight)

Cement: Blast furnace type B   100 kg
       (percentage of stabilizer added: 7.5%)

Separation inhibitor: Acrylic amide type 110 g (powder)
       (82.7 mg/kg, used in a 0.1% solution)

3. *Execution method.* As shown in Photograph 4.1, the execution was performed by supplying the pit sand and cement from a reclaimer outside the bulkhead, mixing these materials on a plant boat (processing capacity of 250 m³/h) inside the bulkhead, and then, after adding the separation inhibitor in the chute boat, placing it in position using the junction chute described in Chapter 6. Figure 4.5 shows the execution process and Photograph 4.2 shows an overall view of the finished work.

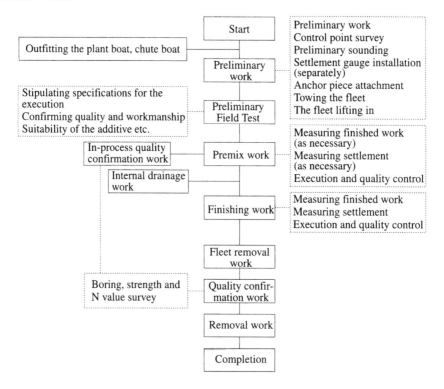

Figure 4.5 Execution procedure.

Photograph 4.1 View of the execution.

Photograph 4.2 After the execution.

4. *Execution management*

4.1 *Material control.* The following methods were used for weight control of all materials: soil, cement, and separation inhibitor.
- Soil: conveyor scale
- Cement: load cell
- Segregation inhibitor solution: flow meter

4.2 *Work progress control.* The execution location and reclamation depth control was performed by the following methods.
- Location control: electro-optical distance meter/goniometer
- Reclamation depth control: automatic sounding lead

4.3 *Quality control.* Tables 4.2 to 4.4 present the quality control items for the landfill materials (soil, cement, separation inhibitor), treated soil, and the ground constructed using the treated soil.

Table 4.2 Landfill material quality control items and details.

Item	Factor Controlled	Control Method	Measurement Frequency
Pit sand	Specific gravity	JIS A 1202 Soil particle density test	Once every 20,000 m^3
	Grain size	JIS A 1204 Soil grain size distribution test (sieve test)	Once every 20,000 m^3
	Water content	JIS A 1203 Water content test	Twice a day
Cement		Test results table	Once at beginning of the work
			Once when the plant or quality changes
			Once a month
Segregation inhibitor		Test results table	Once at beginning of the work
			Once when the plant or quality changes
			Once a month

Table 4.3 Treated soil quality control items and details.

Item	Factor Controlled	Control Method	Measurement Frequency
Mixing control	Calcium	EDTA titration method	Once every 20,000 m^3 (1 sample each time)
	Unconfined compressive strength	JIS A 1216 Unconfined compression test	Once every 20,000 m^3
			Curing days: 1 week, 4 weeks, 13 weeks × 3 specimens
	Density	Same as above. The specimen used for the unconfined compression testing is used	Same as above.
Separation inhibitor solution	Concentration	Obtained from the flow rate of separation inhibitor used and the weight of the separation inhibitor used	Confirmed for each 20,000 m^3 of treated soil

Table 4.4 Quality control items and details of reclamation ground using treated soil.

	Item Surveyed	Survey Method	Measurement Frequency
In-situ	Ground strength	Standard penetration test	Holes at 1 m pitch (boring diameter: 66 mm)
	Elastic wave velocity	Velocity logging (Cross-hole method)	Holes at 1 m pitch (boring diameter: 86 mm) (boring diameter: 66 mm)
Labora-tory *)	Density	Calipers method	1 sample every 2 m
	Unconfined compressive strength	JIS A 1216 Unconfined compression test	1 sample every 2 m (3 specimens/1 sample)
	Cement content	Calcium analysis	1 sample every 2 m
	Cohesion and angle of shear resistance	Triaxial compression test JGS T 523 (Consolidated-undrained test (CU) of soil) JGS T 524 (Consolidated-drained test (CD) of soil)	1 sample every 5 m (4 specimens/1 sample)
	Coefficient of Permeability	Permeability test	1 sample in each hole (performed in 4 holes)

*) Specimens for the laboratory tests are obtained by triple core tube sampling (diameter: 86mm, external diameter: 116mm) or all cores sampling (boring: diameter 116mm).

4.3 NIIGATA AIRPORT

1. *Outline of the work*. This project was performed to expand the runway of the Niigata Airport while it was in use, as shown in Figures 4.6, 4.7 and 4.8 are the plane diagram and section diagram of execution respectively. The landfill material used was sandy soil. It was decided that the soil should be improved because if it were used without improvement, it would be highly likely to liquefy during an earthquake. But because it was a project which involved the extension of a runway while it was in use, a height limitation was imposed on the work site preventing the use of ground improvement methods requiring pile drivers etc., so the premixing method was selected. The capacity of the mixing plant was 250 m^3/h and the material was transported and spread by dump trucks and bulldozers. The design strength of the reclamation ground was the unconfined compressive strength of 100 kN/m^2 (age: 28 days).

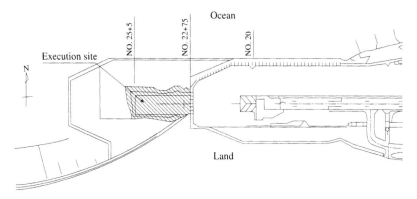

Figure 4.6 Map of the location of the execution.

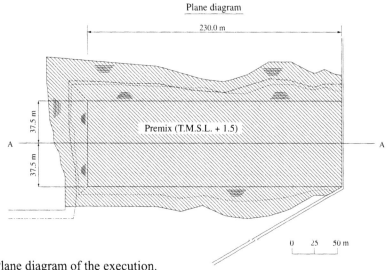

Figure 4.7 Plane diagram of the execution.

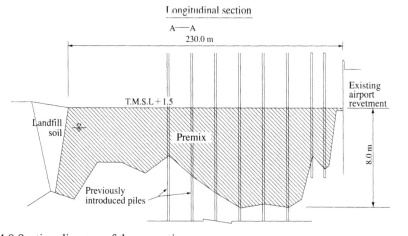

Figure 4.8 Section diagram of the execution.

2. *Quantity of work*

2.1 *Execution quantities*
  - Preliminary field test: 200 m^3
  - Execution: 118,510 m^3

2.2 *Mix proportion at job site.* The following is the mix proportion per 1 m^3 with the required additional rate of 1.7 when the design strength q_{ud} = 100 kN/m^2 (age: 28 days).

  Soil: Dredged and temporary placed soil   1,350 kg (dry weight)
  Cement: Blast furnace type B   40.5 kg
      (percentage of stabilizer added: 3%)
  Separation inhibitor: Acrylic amide, 50.6 liters
      (used in a 0.2% solution)

3. *Execution method.* Photographs 4.3 to 4.6 show the state during execution. After the temporarily placed dredged soil was supplied to the mixing plant by land transport and the treated soil was transported on land by dump trucks, it was spread by bulldozers.

  The execution flow chart is shown in Figure 4.9.

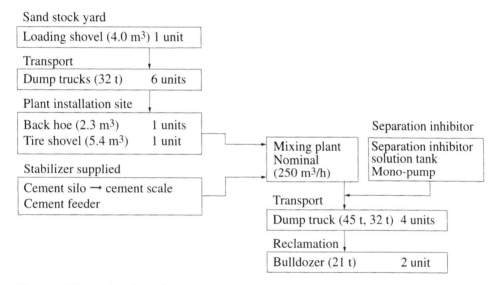

Figure 4.9 Execution flow chart.

Photograph 4.3 View before work commenced.

Photograph 4.4 View of the mixing plant.

Photograph 4.5 Loading on the dump truck.

Photograph 4.6 View of the reclamation.

4. *Execution control*

4.1 *Material control.* The following methods were used for weight control of all materials: soil, cement, and separation inhibitor.

- Soil: conveyor scale
- Cement: load cell
- Separation inhibitor solution: flow meter

4.2 *Finished work control.* It is controlled in the same way as ordinary reclamation work.

4.3 *Quality control.* Table 4.5 and Table 4.6 show the quality control items during the preliminary field test and the successive execution.

Table 4.5 Quality control items (preliminary field test).

Item	Method	Frequency	Quantities
Cement content test	JIS R 5202	10 minute intervals	1/time × 6 times = 6 Sampled from the conveyor
Unconfined compression test	JIS A 1216	10 minute intervals	6 specimens × 18 times = 108 specimens
Wet density test	JGS T 191		6 specimens × 15 times = 90 specimens
Density test (RI) Water content test (RI)	JGS 1614		Untreated soil: 2 times; untreated ground: 1 time; treated ground: 2 times
Maximum and minimum density test	JGS T 161		
Elastic wave velocity measurement (seismic velocity logging)	JGS 1122		
Standard penetration test	JIS A 1219		Before vibration test: 5 times × 6 holes = 30 times After vibration test: 6 times × 2 holes = 12 times
Vibration test			Untreated ground: 3 times; treated ground: 6 times

Table 4.6 Quality control items (execution).

Item	Method	Frequency	Quantities
Cement content test	JIS R 5202	1 time/20,000 m^3	1 specimen/1 time
Unconfined compressive strength test	JIS A 1216	1 time/20,000 m^3	3 specimens × 2 times = 6 specimens (age: 7 days, 28 days)
Water content test	RI	a.m. and p.m. on work days	2 times
Density test Standard penetration test			5 holes (age: 28 days)

4.4 NEW EAST HARBOR IN THE PORT OF ISHIKARI (– 7.5m QUAYWALL)

1. *Outline of the work.* The premixing method was applied to reclamation ground behind an anchor steel sheet pile type quaywall in order to prevent liquefaction at the 7.5 m quaywall shown in Figure 4.10, which is located inside the pier in the New East Harbor of the Port of Ishikari. Figure 4.11 is the plane diagram and Figure 4.12 is the section diagram of the execution. The capacity of the mixing plant was 125 m^3/h and the treated soil was transported and placed at the landfill site by dump trucks and bulldozers. The design strength of the reclaimed land was the unconfined compressive strength of 50 kN/m^2 (age 28 days).

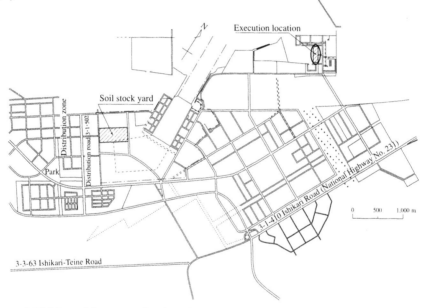

Figure 4.10 Map of the execution location.

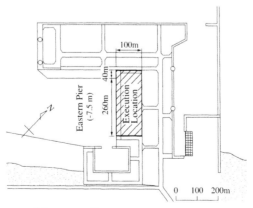

Figure 4.11 Plane diagram of the execution.

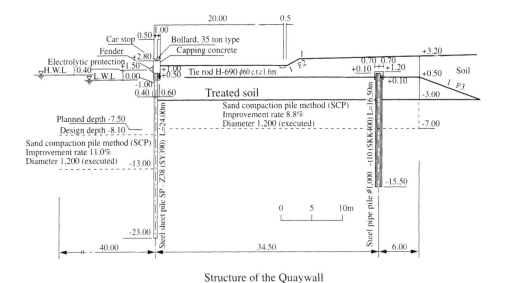

Structure of the Quaywall

Figure 4.12 Section diagram of the execution.

2. *Quantity of work*

2.1 *Execution quantities*

- Preliminary field test: 1,160 m³
- Execution: 57,000 m³

2.2 *Mix proportion at job site.*
The following is the mix proportion per 1 m³ with the required additional rate of 2.0 when the design strength q_{ud} = 50 kN/m² (age: 28 days).

Soil: Dredged and temporary placed soil   1,350 kg (dry weight)

Cement: Blast furnace type B   54 kg
    (percentage of stabilizer added: 4%)

Separation inhibitor: Acrylic amide type   101.25 liters
    (used in a 0.1% solution)

3. *Execution method.*
Photographs 4.7 to 4.11 show the state before and after the execution. After the temporarily placed dredged soil was supplied to the mixing plant by land transport and the treated soil was transported on land by dump trucks, it was spread by bulldozers.

The execution flow chart is shown in Figure 4.13.

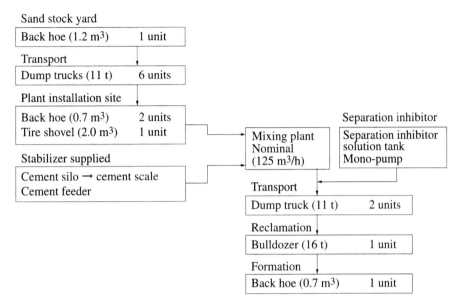

Figure 4.13 Execution flow chart.

Photograph 4.7 View before execution.

Photograph 4.8 View during execution.

Photograph 4.9 View after completion of the execution.

Photograph 4.10 Mixing the treated soil.

Photograph 4.11 Addition of the separation inhibitor.

4. *Execution control*

4.1 *Material control.* The following methods were used for weight control of all materials: soil, cement, and separation inhibitor.

 - Soil: conveyor scale
 - Cement: rotary feeder

- Separation inhibitor solution: flow meter

4.2 *Finished work control.* It is controlled in the same way as ordinary reclamation work.

4.3 *Quality control.* Table 4.7 and Table 4.8 show the quality control items during the preliminary field test and the successive execution.

Table 4.7 Quality control items (preliminary field test).

Item	Method	Frequency	Quantities
Cement content test	JIS R 5202 Hydrochloric acid dissolution method	1 time/h 1 time/h	1 specimen/time × 6 times = 6 specimens 3 specimen/time × 6 times = 18 specimens
Unconfined compression test	JIS A 1216	10:00 a.m.	24 specimens × 2 times = 48 specimens
Wet density test	JGS T 191	3:00 p.m.	
Water content	Infrared moisture meter	1 time/h	
Soil particle density test	JIS A 1204	10:00 a.m.	2 times
Grain size test	JIS A 1202	3:00 p.m.	
Maximum and minimum density test	JGS T 161		
Density test (RI)	JGS 1614	Yard 8:00 a.m. 1:00 p.m. Site 9:00 a.m. 2:00 p.m.	3 specimens/time × 2 times = 6 specimens
		Improved ground 6:00 p.m.	3 specimens/location × 2 locations = 6 specimens

Table 4.8 Quality control items (execution).

Item	Method	Frequency	Quantities
Cement content test	JIS R 5202 Hydrochloric acid dissolution method	1 time/5,000 m^3	3 specimens/1 time
Unconfined compression test	JIS A 1216	1 time/5,000 m^3	3 specimens × 2 times = 6 specimens (age: 7 days, 28 days)
Water content test	Infrared moisture meter	a.m. and p.m. on working days	2 times
Density test (RI)	JGS 1614	1 time/5,000 m^3	3 specimens/1 time
Density test (sand replacement)	JIS A 1214	1 time/5,000 m^3	3 specimens/1 time

4.5 ROKKO ISLAND IN THE PORT OF KOBE (–10 M QUAYWALL)

1. *Outline of the work.* Many harbor structures in the Port of Kobe were damaged by the 1995 Hyogo-ken Nanbu Earthquake. To restore the damaged structures, the restoration method best suited to each structure was selected taking into account its state of deformation and other conditions.

Since this structure (total length = 1,500 m) suffered relatively less deformation than others and its face line remained relatively linear after the deformation, the deformed structure was left unchanged and the restoration method chosen was the premixing method. This permitted the reduction of the earth pressure acting on the quaywall from the ground behind it during earthquakes.

Figures 4.14 and 4.15 show the execution location and the cross section of the execution. The work was performed by excavating and removing the soil from behind the quaywall above the excavation line shown in Figure 4.15. Then it was back filled using soil treated with cement (treated soil). The back fill soil was made of two kinds of soil: high water content soil excavated from behind the quaywall mixed with a 2-shaft mixer and low water content purchased soil mixed by the conveyor method. The design strength of the reclamation ground was the unconfined compressive strength of 100 kN/m^2 (age: 28 days).

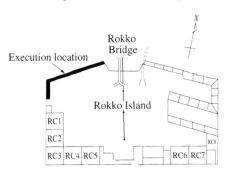

Figure 4.14 Map of the execution location.

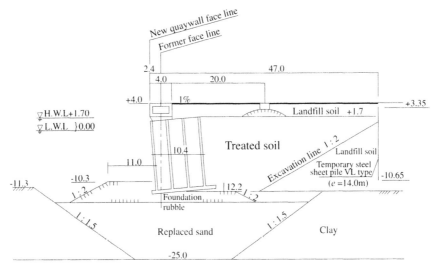

Figure 4.15 Section diagram of the execution.

2. *Quantity of work*

2.1 *Execution quantities*

- Execution: 620,000 m^3

2.2 *Mix proportion at job site.*

The following is the mix proportion per 1 m^3 with the required additional rate of 1.8 when the design strength q_{ud} = 100 kN/m^2 (age: 28 days).

- Excavated soil (grain size: max. 100 mm)
- Purchased soil (grain size: max. 100 mm, produced at Iejima in Hyogo Prefecture)
- Blast furnace cement type B: Percentage of stabilizer added: 7.5%
- Separation inhibitor: 75 mg per 1 kg (dry weight) of the sand

3. *Execution method.*

Two kinds of soil—excavated soil from behind the quaywall and purchased soil—were used as the landfill soil for this construction work. These were mixed separately in two mixing plants, as shown in Figure 4.16. Because the excavated soil has a high water content, material with a grain diameter of 100 mm or less was separated and mixed in a 2-shaft mixer. Because the purchased soil had low water content, it was mixed on a belt conveyor. While the water depth was large, the reclamation was performed by the open-bottom bucket method in order to minimize separation. When it became small, it was placed by dump trucks and bulldozers. Photographs 4.12 to 4.14 show scenes during the execution and Figure 4.17 is a flow chart of the procedure.

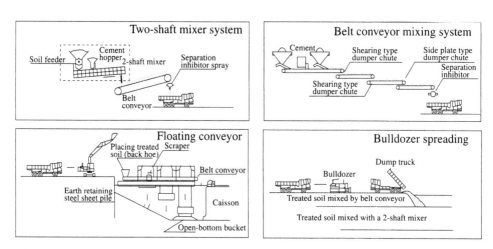

Figure 4.16 Conceptual diagrams of the execution.

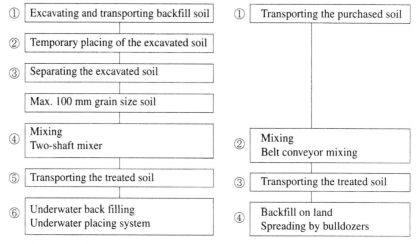

Figure 4.17 Execution flow chart.

Photograph 4.12 View of the soil improvement system.

Photograph 4.13 View of large particle screen system on land.

Photograph 4.14 View of the open bottom bucket mode floating conveyor.

4. *Execution control*

4.1 *Material control*. The following methods were used for weight control of all materials: soil, cement, and separation inhibitor.

- Soil: conveyor scale
- Cement: load cell
- Segregation inhibitor solution: flow meter

4.2 *Finished work control*. It is controlled in the same way as ordinary reclamation work.

4.3 *Quality control*

- Excavated soil and purchased soil
    Density of soil particle
    Grain size distribution
    Water content
- Treated soil

Percentage of cement added
Unconfined compressive strength
- Completed reclamation ground
  Standard penetration test
  Unconfined compressive strength
  Core observations

4.6 PORT OF KOBE, ROKKO ISLAND (FERRY QUAYWALL (RF-3))

1. *Outline of the work*. This structure was constructed using a premixing method that can reduce the earth pressure acting on a quaywall and prevent liquefaction of the landfill soil in order to improve the seismic performance of an existing quaywall. Figures 4.18 to 4.20 show the execution location, plane diagram of the execution, and section diagram of the execution. The execution was performed by excavating and removing soil from behind the quaywall above the excavation line shown in Figure 4.20, and then backfilling with cement treated soil (the premixing method). The design strength of the reclamation ground was the unconfined compressive strength of 100 kN/m^2 (age 28 days).

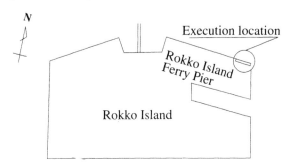

Figure 4.18 Execution location.

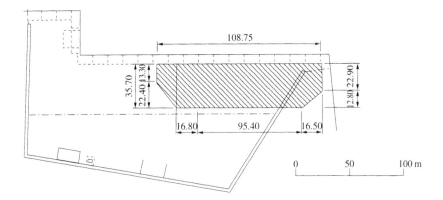

Figure 4.19 Plane diagram of the execution.

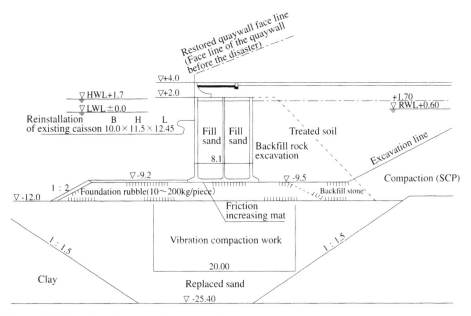

Figure 4.20 Section diagram of the execution.

2. *Quantity of work*

2.1 *Execution quantities*

- Execution: 34,300 m^3

2.2 *Mix proportion at job site.*

The following is the mix proportion per 1 m^3 with the required additional rate of 1.8 when the design strength $q_{ud} = 100$ kN/m^2 (age: 28 days).

Soil: Pit sand   1,400 kg

Cement: Blast furnace type B   105 kg

(percentage of stabilizer added: 7.5%)

Separation inhibitor: Acrylic amide type, 105.0 liters

Used in a 0.1% solution

3. *Execution method.*

This work was divided into 2 steps: initial execution and final execution (see Figure 4.21). The initial execution was performed by placing treated soil on the sea bottom from a gut barge equipped with a grab bucket to form a +1.7 m work area with a lateral size of 10 m behind the central caisson.

The final execution was done by taking treated soil from a mixing plant barge and temporarily placing it in the work area formed by the initial execution, then spreading this soil with a large back hoe (-3 m and deeper) and with a bulldozer (shallower than -3 m).

Figures 4.22 and 4.23 show an outline of the mixing plant and the execution flow chart. The work in progress is shown in Photographs 4.15 to 4.18.

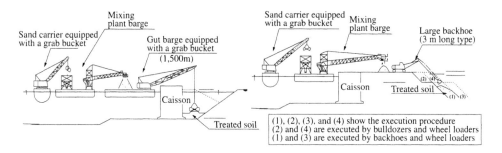

Figure 4.21 Execution outline.

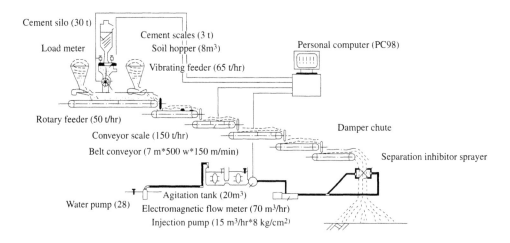

Figure 4.22 Schematic diagram of the soil mixing.

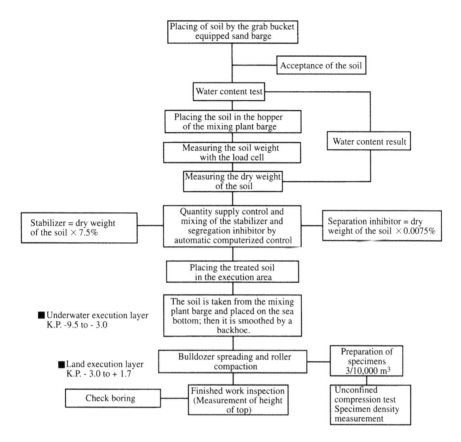

Figure 4.23 Execution flow chart.

Photograph 4.15 Placing the treated soil (below sea surface).

Photograph 4.16 Placing the treated soil (above sea surface).

Photograph 4.17 Placing treated soil and reclamation work.

Photograph 4.18 Placing and spreading treated soil.

4. *Execution control*

4.1 *Material control.* The following methods were used for weight control of all materials: soil, cement, and separation inhibitor.

- Soil: conveyor scale
- Cement: load cell
- Separation inhibitor solution: flow meter

4.2 *Finished work control.* It is controlled in the same way as ordinary reclamation work.

4.3 *Quality control*

- Materials

    Density of soil particle

    Grain size distribution

    Maximum and minimum density

- Mixed soil

    Percentage of cement added

    Unconfined compressive strength

    Wet density

    Water contents

- Treated soil after completion

    Standard penetration test

    Sand sampling

    Unconfined compressive strength

4.7 GROUND SURVEY RESULTS

The following are the results of unconfined compression test of work executed by the premixing method.

1. *Materials used.* The following are the materials used for the work and the properties of the materials.

    Site A: pit sand

    Site B: dredged soil

    Site C: dredged soil

    Site D: decomposed granite soil (excavated soil previously used for reclamation)

Table 4.9 Physical properties of the materials used.

	Site A	Site B	Site C	Site D
Density of the soil particle ρ_s	2.715	2.664	2.690	2.690
Natural water content w (%)	6.5	9.4	11.8	
Uniformity coefficient U_c	1.9	1.8	2.6	55.4
Max. grain size (mm)	4.75	0.85	2.00	100.0
Fines content P (%)	3.35	0	2.00	11.7
Max. density ρ_{dmax} (g/cm^3)	1.580	1.562	1.599	
Min. density ρ_{dmin} (g/cm^3)	1.302	1.251	1.242	

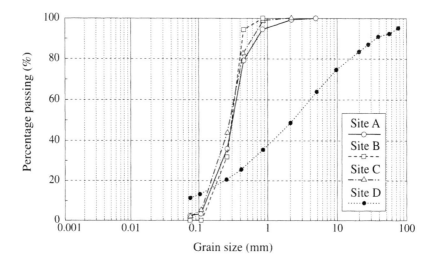

Figure 4.24 Grain size distribution of the materials used.

2. *Design strength.* The following are the design strengths for each site.

   Site A: Unconfined compressive strength q_{ud}
   = 400 kN/m^2 age: 91 days
   Site B: Unconfined compressive strength q_{ud}
   = 100 kN/m^2 age: 28 days
   Site C: Unconfined compressive strength q_{ud}
   = 50 kN/m^2 age: 28 days
   Site D: Unconfined compressive strength q_{ud}
   = 100 kN/m^2 age: 28 days

3. *Unconfined compression testing results*

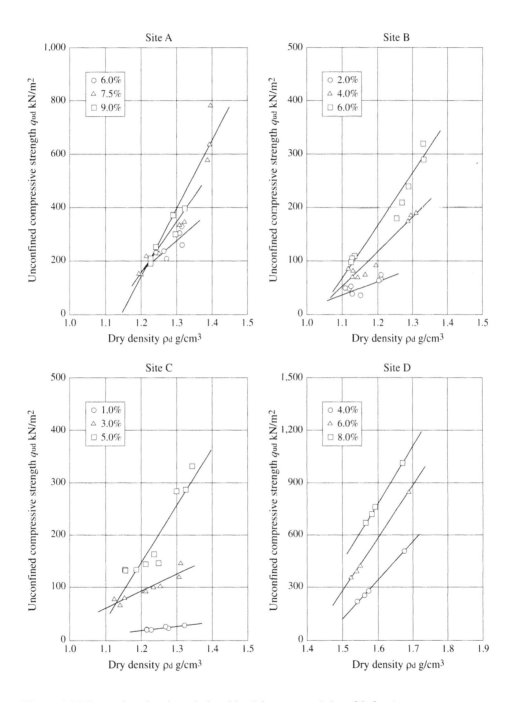

Figure 4.25 Strength – density relationship (laboratory mixing, 28 days).

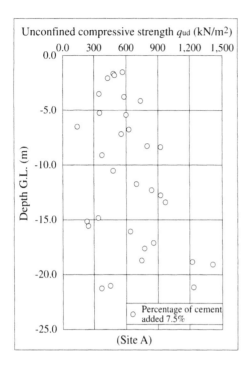

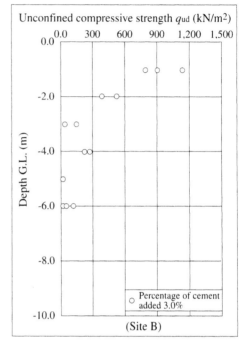

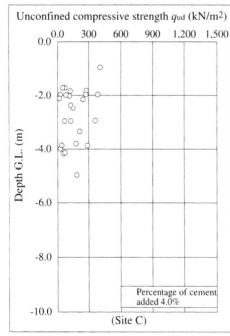

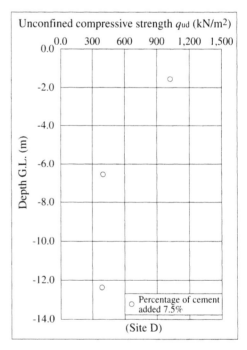

Figure 4.26 Unconfined compressive strength – depth relationship (field).

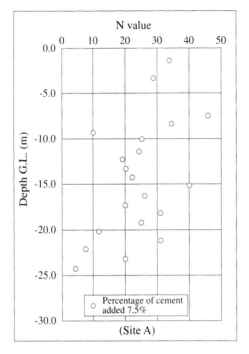

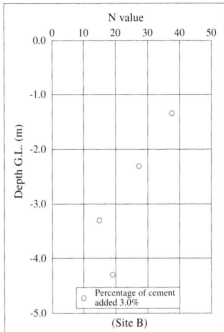

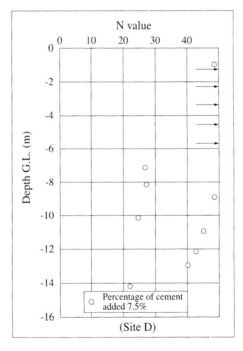

Figure 4.27 N value – depth relationship.

CHAPTER 5

Design Method

5.1 BASIC CONCEPTS

The design will determine the required strength of the treated soil and the improvement range.

[Commentary]
1. The design strength and the improvement range of the treated soil is determined in line with the procedure in Figure 5.1. After the design strength has been determined, the mix proportion design shown in Figure 5.4 is done to set the percentage of stabilizer added to the soil. A separate study must be performed to design the structural details of the caisson body etc.
2. When the method is to be used to prevent liquefaction, the percentage of stabilizer added should be adequate to provide sufficient cohesion to prevent liquefaction.
3. The evaluation of the earth pressure reduction effect or the study of the ground stability against circular slip, must consider the treated soil to be a c - ϕ material. The treated soil is considered to be a c - ϕ material because the shear strength of a treated soil made of a sandy parent soil is represented by Equation 5.

$$\tau_f = c_d + \sigma' \, tan \, \phi_d \, (kN/m^2) \quad \cdots\cdots\cdots\cdots\cdots\cdots\cdots\cdots\cdots\cdots\cdots\cdots\cdots \quad (5.1)$$

Where:

τ_f: shear strength of the treated soil

σ' : effective normal stress

c_d, ϕ_d: cohesion and angle of shear resistance obtained from consolidated drained triaxial compression test

4. The earth pressure acting on the wall surface from the treated ground is calculated by the method in Section 5.2.3.
5. The improvement range is determined by studying stability on sliding, overturning, bearing failure. etc. of the structure. And, because it is assumed that the treated ground has a rigidity greater than that of the untreated ground

surrounding it and behaves as a rigid body, the overall sliding stability including the structure is studied.

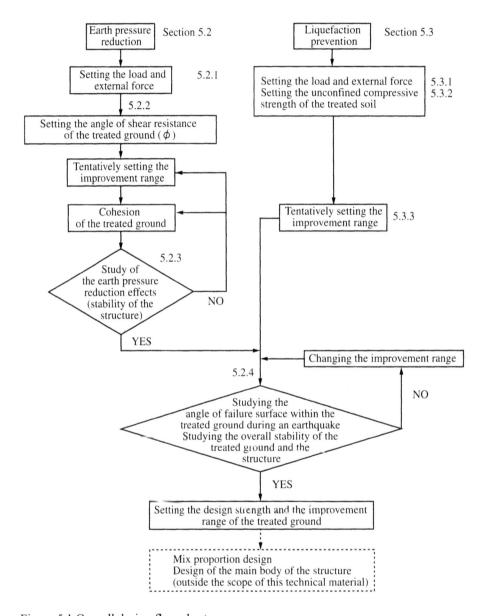

Figure 5.1 Overall design flow chart.

5.2 DESIGN TO REDUCE EARTH PRESSURE

The design is performed as in Figure 5.2 when the method is to be used to reduce earth pressure.

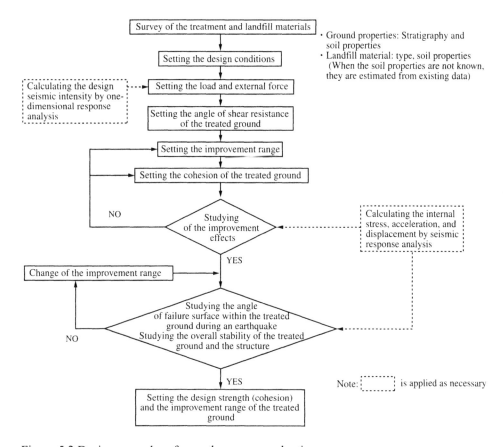

Figure 5.2 Design procedure for earth pressure reduction.

5.2.1 *External forces*
The external forces used for the design of the treated ground are calculated by appropriate methods.

[Commentary]
1. The major loads and external forces that must be accounted for are the surcharge, self weight of the treated ground, buoyancy, earth pressure, residual water pressure, reaction force of fender, seismic force, dynamic water pressure, and wave force, etc.
2. Seismic external forces such as earth pressure during an earthquake are calculated by the seismic coefficient method. The apparent seismic coefficient

below the water used in this case is obtained by using the equation of Arai and Yokoi shown as Equation (5.2).[1] Consequently, the designer must account for the dynamic water pressure that will act on the front surface of the wall body. To do so, the direction of the action of the dynamic water pressure conforms to the inertia direction on the wall body.

$$k' = \frac{2\,(\Sigma \gamma_t\, h_i + \Sigma \gamma\, h_j + \omega) + \gamma h}{2\,\{\Sigma \gamma_t\, h_i + \Sigma\,(\gamma - 10)\, h_j + \omega\} + (\gamma - 10)\, h}\ k \quad\cdots\cdots\cdots\cdots\cdots \quad (5.2)$$

Where:

k': apparent seismic coefficient

γ_t: unit weight of soil above the residual water level (kN/m³)

h_i: thickness of the i-th soil layer above the residual water level (m)

γ: saturated unit weight of soil below the residual water level (kN/m³)

h_j: thickness of the j-th soil layer below the residual water level (m)

ω: surcharge per unit area at the ground surface (kN/m³)

h: thickness of soil layer where the earth pressure is calculated below the residual water level (m)

k: seismic coefficient

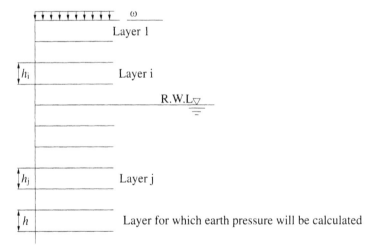

Figure 5.3 Symbols indicating apparent seismic coefficient.

3. The dynamic water pressure is calculated by Equation (5.3).

$$P_{dw} = \pm 7/8\, k\gamma_w\, \sqrt{\mathrm{Hy}} \quad\cdots\cdots\cdots\cdots\cdots\cdots\cdots\cdots\cdots\cdots\cdots\cdots\cdots\cdots\cdots \quad (5.3)$$

Where

P_{dw}: dynamic water pressure (kN/m²)

k: seismic coefficient

γ_w: unit weight of water (kN/m³)

y: depth from the still water level (m)

H: height of structure below the still water level (m)

The resultant force of the dynamic water pressure and its acting depth shall be calculated by the following equation.

$$P_{dw} = \pm 7/12 \, k\gamma_w H^2 \, (= \int_0^H 7/8 \, k\gamma_w \sqrt{H} \, dy) \quad \cdots\cdots\cdots\cdots\cdots\cdots\cdots\cdots\cdots\cdots\cdots \quad (5.4)$$
$$H_{dw} = 3H/5$$

Where

P_{dw}: resultant force of the dynamic water pressure (kN/m)

H_{dw}: depth of the acting point of the resultant force from the still water level (m)

The design seismic coefficient is usually calculated from the regional seismic coefficient, subsoil condition factor, and importance factor. Another method is to obtain the design seismic coefficient by the one-dimensional seismic response calculation method. In the latter method, the response acceleration at the ground surface is calculated from the expected bedrock motion and is converted to the design seismic coefficient by Noda and Uwabe's equation. Reference document (Arai & Yokoi, 1965) gives details of this method and reference document (Coastal Development Institute of Technology, 1997) describes the modeling of the ground for one-dimensional seismic response analysis. The shear modulus of the treated soil used as the input parameter is, as shown in 3.2 (2.3.1) Shear modulus, greater than that of the untreated soil. The detailed values of shear modulus are set appropriately by laboratory test.

5.2.2 Setting the strength of the treated soil

The strength of the treated soil is set accounting for the application and the application conditions of the method so that the required improvement effects are obtained. But the improvement range (depth) must be assumed when the design sections, such as a section of the structure etc., are not set in advance.

[Commentary]

The strength of the treated soil is set based on the flow chart in Figure 5.4.

1. As shown in Equation (5.1), the strength of the treated soil is estimated from *c*, ϕ, and the effective normal stress. Among these, the cohesion *c* is highly dependent on the percentage of stabilizer added, which can be controlled by the premixing method. The factor represented by ϕ is not as influenced by the percentage of stabilizer added as cohesion *c*. Consequently, setting the strength of the treated soil is equivalent to setting the cohesion *c*.

2. The safety factors used for the study of stability against sliding, overturning, and bearing failure etc. of the structure and the bottom friction coefficient of the caisson used were those of the structures used in the past.

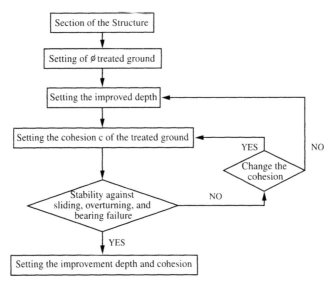

Figure 5.4 Treated soil strength setting flow chart.

3. According to past consolidated drained triaxial compression tests of treated soil prepared with a percentage of added stabilizer of 10% or less, the angle of shear resistance of the treated soil is almost equal to or slightly greater than that of the parent soil, as shown in Figure 5.5. Consequently, for design purposes, the angle of shear resistance of the treated soil can be equal to that of the untreated soil for conservative design.

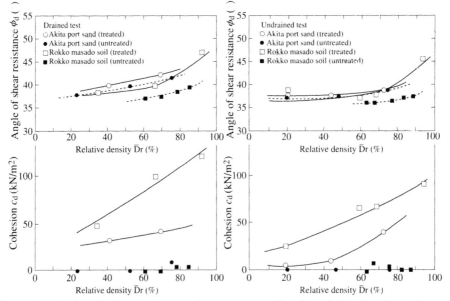

Figure 5.5 Relationship between strength parameter and to the relative density at the peak strength of untreated soil and treated soil.

4. To find the angle of shear resistance from a triaxial test, both the estimated density and effective overburden pressure of the ground after reclamation are obtained from a consolidated drained triaxial test. The angle of shear resistance used for design is generally about 5° to 10° lower than the value obtained by consolidated drained triaxial test. According to Technical Standards and Commentaries for Port and Harbour Facilities in Japan (hereafter abbreviated to TCPHFJ), ϕ_d obtained by the following method can be used as a design constant for stability analysis. With this method, ϕ_d must be obtained by consolidated drained triaxial test on undisturbed in-situ specimens and at a confining pressure that is suited for design conditions.

If ϕ_d is used unchanged, however, the bearing capacity may be overestimated, and so ϕ_d unchanged must be used with considerable caution. In cases where no triaxial test is performed, ϕ may be obtained from the N value of the ground estimated after reclamation (with care taken to use the N value for untreated ground).

5.2.3 *Earth pressure*
The earth pressure acting on a quaywall from an improved body is calculated by an equation that accounts for c - ϕ.

[Commentary]
The earth pressure of treated soil can be calculated by the following equations.

$$p_{ai} = \left\{ \frac{(\Sigma \gamma_i \cdot h_i)\ \cos(\psi - \beta)}{\cos \psi} + \omega \right\} \frac{\sin(\zeta_i - \phi_i + \theta)\ \cos(\psi - \zeta_i)}{\cos \theta\ \cos(\psi - \zeta_i + \phi_i + \delta)\ \sin(\zeta_i - \beta)}$$

$$- \frac{c_i\ \cos(\psi - \beta)\ \cos \phi_i}{\cos(\psi - \zeta_i + \phi_i + \delta)\ \sin(\zeta_i - \beta)} \quad\dots\dots\dots\dots\dots\dots\dots\dots \quad (5.5)$$

$$2\zeta_i = \psi + \phi_i - \mu_i + 90°$$

$$\mu_i = tan^{-1} \frac{B_i C_i + A_i\ \sqrt{B_i^2 - A_i^2 + C_i^2}}{B_i^2 - A_i^2}$$

$$A_i = sin(\delta + \beta + \theta)$$

$$B_i = sin(\psi + \phi_i + \delta - \beta)\ \cos\theta - sin(\psi - \phi_i + \theta)\ \cos(\delta + \beta)$$

$$+ \frac{2c_i\ \cos(\psi - \beta)\ \cos \phi_i\ \cos(\delta + \beta)\ \cos\theta}{\dfrac{(\Sigma \gamma_i \cdot h_i)\ \cos(\psi - \beta)}{2 \cos \psi} + \omega}$$

$$C_i = sin(\psi + \phi_i + \delta - \beta)\ \sin\theta + sin(\psi - \phi_i + \theta)\ \sin(\delta + \beta)$$

$$- \frac{2c_i\ \cos(\psi - \beta)\ \cos \phi_i\ \sin(\delta + \beta)\ \cos\theta}{\dfrac{(\Sigma \gamma_i \cdot h_i)\ \cos(\psi - \beta)}{2 \cos \psi} + \omega}$$

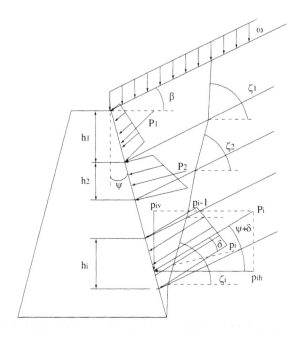

Figure 5.6 Earth pressure during an earthquake.

Where:

p_{ai}: active earth pressure acting on the wall surface at the bottom level of the *i*-th soil layer (kN/m²)

c_i: cohesion of soil in the *i*-th layer (kN/ m²)

ϕ_i: angle of shear resistance of the *i*-th soil layer (°)

γ_i: unit weight of the *i*-th soil layer (kN /m³)

h_i: thickness of the *i*-th soil layer (m)

ψ: angle of backface wall to the vertical (°)

β: angle of backfill ground surface to the horizontal (°)

δ: angle of wall friction (°)

ζ_i: angle of failure surface of the *i*-th soil layer to the horizontal (°)

ω: surcharge per unit surface area of the ground surface (kN/m²)

θ: composite seismic angle which is defined as an angle given by the following equations $\theta = tan^{-1} k$ (°) or it is represented as $\theta = tan^{-1} k'$

k: seismic coefficient

k': apparent seismic coefficient

If $\beta = 0$ and $\psi = 0$, Equation (5.5) can be rewritten as follows.

$$p_{ai} = (\Sigma \gamma_i \cdot h_i + \omega) \frac{sin (\zeta_i - \phi_i + \theta) \cos (-\zeta_i)}{cos\theta \; \cos(-\zeta_i + \phi_i + \delta) \; sin\zeta_i} - \frac{c_i \cos\phi_i}{cos(-\zeta_i + \phi_i + \delta) \; sin\zeta_i}$$

And when:

$$p_{a1} = (\Sigma \gamma_i \cdot h_i + \omega) \frac{sin(\zeta_i - \phi_i + \theta) \, cos(-\zeta_i)}{cos\theta \, cos(-\zeta_i + \phi_i + \delta) \, sin\zeta_i}$$

$$p_{a2} = \frac{c_i \, cos\phi_i}{cos(-\zeta_i + \phi_i + \delta) \, sin\zeta_i}$$

the results of calculating p_{a1}, p_{a2}, and ζ_a under the following conditions are listed in Tables 5.1(a) to (d).

$\Sigma \gamma_i \cdot h_i = 50, 100 \text{kN/m}^2$
$\omega = 0 \text{kN/m}^2$
$\phi = 20, 25, 30, 35°$
$c = 0.0, 50.0, 100 \text{kN/m}^2$
$k = 0.00 \sim 0.40 \ (0.05/\text{device})$
$\delta = +15, 0, -15°$

Table 5.1(a) Active earth pressure and angle of failure surface ($\gamma \cdot h = 50$ kN/m^2).

$\phi = 20°$, $\delta = 0°$

k	c=0.0kN/m^2			c=50kN/m^2			c=100kN/m^2		
	P_{a1}	P_{a2}	ζ_a	P_{a1}	P_{a2}	ζ_a	P_{a1}	P_{a2}	ζ_a
0.00	24.5	0.0	55.0	24.5	70.0	55.0	24.5	140.0	55.0
0.05	26.4	0.0	52.0	26.3	70.0	54.6	26.3	140.0	54.8
0.10	28.4	0.0	48.5	28.1	70.0	54.1	28.1	140.1	54.5
0.15	30.8	0.0	44.3	30.0	70.1	53.7	29.9	140.1	54.3
0.20	33.6	0.0	39.3	32.0	70.1	53.2	31.8	140.1	54.0
0.25	37.0	0.0	33.0	34.0	70.2	52.7	33.6	140.1	53.8
0.30	41.5	0.0	24.7	36.0	70.3	52.2	35.6	140.2	53.5
0.35	49.3	0.0	11.2	38.2	70.4	51.7	37.5	140.2	53.3
0.40				40.4	70.5	51.2	39.5	140.3	53.0

$\delta = +15°$

k	c=0.0kN/m^2			c=50kN/m^2			c=100kN/m^2		
	P_{a1}	P_{a2}	ζ_a	P_{a1}	P_{a2}	ζ_a	P_{a1}	P_{a2}	ζ_a
0.00	21.7	0.0	49.5	20.5	59.9	60.0	20.2	119.5	61.1
0.05	23.8	0.0	46.1	21.8	59.9	59.5	21.4	119.6	60.8
0.10	26.1	0.0	42.3	23.3	60.0	59.0	22.7	119.6	60.6
0.15	28.9	0.0	37.9	24.7	60.1	58.4	24.0	119.7	60.3
0.20	32.3	0.0	32.8	26.3	60.2	57.9	25.3	119.7	60.0
0.25	36.7	0.0	26.8	27.9	60.3	57.4	26.7	119.8	59.7
0.30	42.7	0.0	19.3	29.6	60.5	56.8	28.1	119.9	59.4
0.35	53.7	0.0	8.3	31.3	60.6	56.2	29.5	120.0	59.2
0.40				33.2	60.8	55.6	31.0	120.1	58.9

$\delta = -15°$

k	c=0.0kN/m^2			c=50kN/m^2			c=100kN/m^2		
	P_{a1}	P_{a2}	ζ_a	P_{a1}	P_{a2}	ζ_a	P_{a1}	P_{a2}	ζ_a
0.00	33.1	0.0	67.7	29.5	86.6	49.4	29.1	173.0	48.5
0.05	34.6	0.0	65.7	32.0	86.6	49.1	31.7	172.9	48.3
0.10	36.3	0.0	63.2	34.5	86.5	48.7	34.3	172.9	48.1
0.15	38.1	0.0	60.1	37.1	86.5	48.4	37.0	172.9	48.0
0.20	40.2	0.0	56.0	39.8	86.4	48.0	39.7	171.9	47.8
0.25	42.5	0.0	50.2	42.5	86.4	47.6	42.5	172.9	47.6
0.30	45.4	0.0	41.1	45.2	86.4	47.3	45.2	172.9	47.4
0.35	49.8	0.0	21.3	48.1	86.5	46.9	48.0	172.9	47.2
0.40				51.0	86.5	46.5	50.9	172.9	47.0

$\phi = 25°$, $\delta = 0°$

k	c=0.0kN/m^2			c=50kN/m^2			c=100kN/m^2		
	P_{a1}	P_{a2}	ζ_a	P_{a1}	P_{a2}	ζ_a	P_{a1}	P_{a2}	ζ_a
0.00	20.3	0.0	57.5	20.3	63.7	57.5	20.3	127.4	57.5
0.05	22.0	0.0	55.0	21.9	63.7	57.0	21.9	127.4	57.2
0.10	23.8	0.0	52.1	23.6	63.7	56.6	23.5	127.4	57.0
0.15	25.9	0.0	48.9	25.3	63.8	56.1	25.2	127.4	56.7
0.20	28.2	0.0	45.2	27.1	63.8	55.6	26.9	127.5	56.5
0.25	30.9	0.0	40.9	28.9	63.9	55.1	28.6	127.5	56.2
0.30	34.0	0.0	35.9	30.9	63.9	54.6	30.4	127.5	55.9
0.35	37.9	0.0	29.8	32.8	64.0	54.1	32.2	127.6	55.7
0.40	43.0	0.0	22.1	34.9	64.1	53.5	34.0	127.7	55.4

$\delta = +15°$

k	c=0.0kN/m^2			c=50kN/m^2			c=100kN/m^2		
	P_{a1}	P_{a2}	ζ_a	P_{a1}	P_{a2}	ζ_a	P_{a1}	P_{a2}	ζ_a
0.00	18.2	0.0	53.4	17.2	55.3	62.4	16.9	110.4	63.5
0.05	19.9	0.0	50.5	18.4	55.4	61.9	18.1	110.5	63.2
0.10	22.0	0.0	47.3	19.8	55.4	61.3	19.2	110.5	63.0
0.15	24.3	0.0	43.8	21.1	55.5	60.8	20.4	110.6	62.7
0.20	26.9	0.0	39.9	22.6	55.6	60.2	21.7	110.6	62.4
0.25	30.1	0.0	35.4	24.1	55.8	59.7	22.9	110.7	62.1
0.30	33.9	0.0	30.4	25.6	55.9	59.1	24.3	110.8	61.8
0.35	38.9	0.0	24.7	27.3	56.1	58.5	25.6	110.9	61.5
0.40	45.7	0.0	17.7	29.0	56.2	57.9	27.0	111.0	61.1

$\delta = -15°$

k	c=0.0kN/m^2			c=50kN/m^2			c=100kN/m^2		
	P_{a1}	P_{a2}	ζ_a	P_{a1}	P_{a2}	ζ_a	P_{a1}	P_{a2}	ζ_a
0.00	26.1	0.0	65.1	23.9	77.4	52.1	23.6	154.5	51.1
0.05	27.7	0.0	63.2	26.1	77.3	51.7	25.9	154.5	50.9
0.10	29.4	0.0	61.0	28.4	77.3	51.3	28.3	154.5	50.7
0.15	31.3	0.0	58.4	30.7	77.3	50.9	30.7	154.5	50.5
0.20	33.3	0.0	55.3	33.1	77.2	50.5	33.1	154.4	50.3
0.25	35.6	0.0	51.6	35.6	77.2	50.1	35.6	154.4	50.1
0.30	38.1	0.0	46.9	38.0	77.2	49.7	38.0	154.4	49.9
0.35	41.1	0.0	40.7	40.6	77.2	49.3	40.6	154.5	49.6
0.40	44.8	0.0	31.9	43.2	77.3	48.9	43.1	154.5	49.4

Table 5.1(b) Active earth pressure and angle of failure ($\gamma h = 50$ kN/m^2).

$\phi = 30°$, $\delta = 0°$

k	c=0.0kN/m^2			c=50kN/m^2			c=100kN/m^2		
	P_{a1}	P_{a2}	ζ_a	P_{a1}	P_{a2}	ζ_a	P_{a1}	P_{a2}	ζ_a
0.00	16.7	0.0	60.0	16.7	57.7	60.0	16.7	115.5	60.0
0.05	18.2	0.0	57.8	18.1	57.7	59.5	18.1	115.5	59.7
0.10	19.8	0.0	55.3	19.7	57.8	59.0	19.6	115.5	59.5
0.15	21.6	0.0	52.6	21.2	57.8	58.5	21.1	115.5	59.2
0.20	23.7	0.0	49.6	22.9	57.8	58.0	22.7	115.5	58.9
0.25	25.9	0.0	46.3	24.6	57.9	57.5	24.3	115.6	58.6
0.30	28.5	0.0	42.6	26.3	58.0	56.9	25.9	115.6	58.3
0.35	31.4	0.0	38.4	28.1	58.0	56.4	27.5	115.7	58.0
0.40	34.8	0.0	33.6	30.0	58.1	55.8	29.2	115.7	57.7

$\delta = +15°$

k	c=0.0kN/m^2			c=50kN/m^2			c=100kN/m^2		
	P_{a1}	P_{a2}	ζ_a	P_{a1}	P_{a2}	ζ_a	P_{a1}	P_{a2}	ζ_a
0.00	15.1	0.0	56.9	14.3	50.9	64.8	14.0	101.5	65.9
0.05	16.6	0.0	54.3	15.4	50.9	64.2	15.1	101.6	65.6
0.10	18.4	0.0	51.6	16.7	51.0	63.8	16.2	101.6	65.3
0.15	20.4	0.0	48.6	17.9	51.1	63.1	17.2	101.7	65.0
0.20	22.6	0.0	45.3	19.3	51.2	62.5	18.4	101.8	64.7
0.25	25.2	0.0	41.8	20.7	51.3	61.9	19.6	101.8	64.4
0.30	28.1	0.0	37.8	22.1	51.4	61.3	20.8	101.9	64.1
0.35	31.6	0.0	33.6	23.7	51.6	60.7	22.1	102.0	63.7
0.40	35.9	0.0	28.8	25.2	51.7	60.1	23.3	102.1	63.4

$\delta = -15°$

k	c=0.0kN/m^2			c=50kN/m^2			c=100kN/m^2		
	P_{a1}	P_{a2}	ζ_a	P_{a1}	P_{a2}	ζ_a	P_{a1}	P_{a2}	ζ_a
0.00	20.8	0.0	65.1	19.2	69.0	54.7	18.9	137.7	53.7
0.05	22.3	0.0	63.3	21.2	68.9	54.3	21.0	137.7	53.5
0.10	24.0	0.0	61.3	23.3	68.9	53.9	23.1	137.6	53.3
0.15	25.7	0.0	59.0	25.4	68.8	53.5	25.3	137.6	53.0
0.20	27.6	0.0	56.5	27.5	68.8	53.1	27.5	137.6	52.8
0.25	29.7	0.0	53.6	29.7	68.8	52.7	29.7	137.6	52.6
0.30	32.0	0.0	50.3	31.9	68.8	52.2	31.9	137.6	52.3
0.35	34.5	0.0	46.4	34.2	68.8	51.8	34.2	137.6	52.1
0.40	37.3	0.0	41.7	36.6	68.8	51.3	36.5	137.6	51.9

$\phi = 35°$, $\delta = 0°$

k	c=0.0kN/m^2			c=50kN/m^2			c=100kN/m^2		
	P_{a1}	P_{a2}	ζ_a	P_{a1}	P_{a2}	ζ_a	P_{a1}	P_{a2}	ζ_a
0.00	13.6	0.0	62.5	13.6	52.1	62.5	13.6	104.1	62.5
0.05	14.9	0.0	60.5	14.9	52.1	62.0	14.9	104.1	62.2
0.10	16.4	0.0	58.3	16.3	52.1	61.5	16.2	104.1	61.9
0.15	18.0	0.0	55.9	17.7	52.1	60.9	17.6	104.1	61.6
0.20	19.8	0.0	53.3	19.2	52.1	60.4	19.0	104.2	61.3
0.25	21.7	0.0	50.6	20.7	52.2	59.8	20.4	104.2	61.0
0.30	23.3	0.0	47.6	22.3	52.3	59.3	21.9	104.2	60.7
0.35	26.3	0.0	44.3	24.0	52.4	58.7	23.4	104.3	60.4
0.40	29.1	0.0	40.7	25.7	52.4	58.1	25.0	104.4	60.1

$\delta = +15°$

k	c=0.0kN/m^2			c=50kN/m^2			c=100kN/m^2		
	P_{a1}	P_{a2}	ζ_a	P_{a1}	P_{a2}	ζ_a	P_{a1}	P_{a2}	ζ_a
0.00	12.4	0.0	60.1	11.7	46.5	67.1	11.5	92.9	68.3
0.05	13.8	0.0	57.8	12.8	46.0	66.6	12.5	92.9	68.0
0.10	15.3	0.0	55.4	13.9	46.0	66.0	13.4	92.9	67.7
0.15	17.0	0.0	52.7	15.1	46.7	65.4	14.5	93.0	67.4
0.20	19.0	0.0	49.9	16.3	46.8	64.8	15.5	93.1	67.0
0.25	21.1	0.0	46.9	17.6	46.9	64.2	16.6	93.1	66.7
0.30	23.6	0.0	43.6	19.0	47.1	63.5	17.7	93.2	66.3
0.35	26.4	0.0	40.1	20.4	47.2	62.9	18.9	93.3	66.0
0.40	29.6	0.0	36.4	21.9	47.4	62.2	20.1	93.4	65.6

$\delta = -15°$

k	c=0.0kN/m^2			c=50kN/m^2			c=100kN/m^2		
	P_{a1}	P_{a2}	ζ_a	P_{a1}	P_{a2}	ζ_a	P_{a1}	P_{a2}	ζ_a
0.00	16.5	0.0	66.1	15.3	61.2	57.4	15.0	122.2	56.3
0.05	17.9	0.0	64.4	17.1	61.1	57.0	16.9	122.1	56.1
0.10	19.4	0.0	62.5	18.9	61.1	56.5	18.8	122.1	55.8
0.15	21.1	0.0	60.5	20.8	61.1	56.1	20.7	122.1	55.6
0.20	22.8	0.0	58.3	22.7	61.1	55.6	22.7	122.1	55.3
0.25	24.7	0.0	55.9	24.7	61.0	55.2	24.7	122.1	55.1
0.30	26.7	0.0	53.3	26.7	61.0	54.7	26.7	122.1	54.8
0.35	28.9	0.0	50.3	28.7	61.1	54.2	28.7	122.1	54.6
0.40	31.4	0.0	47.0	30.9	61.1	53.7	30.8	122.1	54.3

Table 5.1(c) Active earth pressure and angle of failure surface ($\gamma h = 100$ kN/m²).

$\phi = 20°$, $\delta = 0°$

k	c=0.0kN/m²			c=50kN/m²			c=100kN/m²		
	Pa1	Pa2	ζa	Pa1	Pa2	ζa	Pa1	Pa2	ζa
0.00	49.0	0.0	55.0	49.0	70.0	55.0	49.0	140.0	55.0
0.05	52.7	0.0	52.0	52.6	70.0	54.2	52.6	140.1	54.6
0.10	56.9	0.0	48.5	52.6	70.0	54.2	52.6	140.1	54.6
0.15	61.6	0.0	44.3	60.4	70.2	52.6	60.0	140.2	53.7
0.20	67.2	0.0	39.3	64.5	70.3	51.8	63.9	140.2	53.2
0.25	74.0	0.0	33.0	69.0	70.6	50.9	67.9	140.3	52.7
0.30	83.1	0.0	24.7	73.7	70.8	50.0	72.1	140.5	52.2
0.35	98.6	0.0	11.2	78.6	71.2	49.0	76.4	140.7	51.7
0.40				83.9	71.6	48.0	80.8	141.0	51.2

$\delta = +15°$

k	c=0.0kN/m²			c=50kN/m²			c=100kN/m²		
	Pa1	Pa2	ζa	Pa1	Pa2	ζa	Pa1	Pa2	ζa
0.00	43.4	0.0	49.5	41.7	60.1	58.3	41.0	119.7	60.0
0.05	47.5	0.0	46.1	44.8	60.3	57.4	43.7	119.9	59.5
0.10	52.3	0.0	42.2	48.0	60.6	56.5	46.5	120.0	59.0
0.15	57.9	0.0	37.9	51.5	60.9	55.5	49.5	120.2	58.4
0.20	64.7	0.0	32.8	55.3	61.2	54.5	52.6	120.4	57.9
0.25	73.3	0.0	26.8	59.3	61.6	53.5	55.8	120.7	57.4
0.30	85.4	0.0	19.3	63.7	62.1	52.4	59.2	121.0	56.8
0.35	107.4	0.0	8.3	68.3	62.7	51.3	62.7	121.3	56.2
0.40				73.4	63.4	50.1	66.4	121.7	55.6

$\delta = -15°$

k	c=0.0kN/m²			c=50kN/m²			c=100kN/m²		
	Pa1	Pa2	ζa	Pa1	Pa2	ζa	Pa1	Pa2	ζa
0.00	66.1	0.0	67.7	60.0	87.0	51.0	58.9	173.2	49.4
0.05	69.3	0.0	65.7	64.7	86.8	50.4	63.9	173.1	49.1
0.10	72.6	0.0	63.2	69.5	86.7	49.8	69.0	173.0	48.7
0.15	76.3	0.0	60.1	74.5	86.6	49.1	74.2	172.9	48.4
0.20	80.3	0.0	56.0	79.6	86.5	48.5	79.5	172.9	48.0
0.25	85.0	0.0	50.2	85.0	86.4	47.8	85.0	172.9	47.6
0.30	90.8	0.0	41.1	90.5	86.4	47.0	90.5	172.9	47.3
0.35	99.6	0.0	21.3	96.2	86.5	46.3	96.1	172.9	46.9
0.40				102.3	86.6	45.4	101.9	173.0	46.5

$\phi = 25°$, $\delta = 0°$

k	c=0.0kN/m²			c=50kN/m²			c=100kN/m²		
	Pa1	Pa2	ζa	Pa1	Pa2	ζa	Pa1	Pa2	ζa
0.00	40.6	0.0	57.5	40.6	63.7	57.5	40.6	127.4	57.5
0.05	43.9	0.0	55.0	43.9	63.7	56.7	43.8	127.4	57.0
0.10	47.6	0.0	52.1	47.3	63.8	55.9	47.2	127.5	56.6
0.15	51.7	0.0	48.9	50.9	63.9	55.1	50.6	127.5	56.1
0.20	56.4	0.0	45.2	54.7	64.0	54.2	54.1	127.6	55.6
0.25	61.7	0.0	40.9	58.8	64.2	53.3	57.9	127.7	55.1
0.30	68.0	0.0	35.9	63.0	64.4	52.4	61.7	127.9	54.6
0.35	75.8	0.0	29.8	67.6	64.7	51.4	65.7	128.1	54.1
0.40	86.0	0.0	22.1	72.4	65.1	50.4	69.7	128.3	53.5

$\delta = +15°$

k	c=0.0kN/m²			c=50kN/m²			c=100kN/m²		
	Pa1	Pa2	ζa	Pa1	Pa2	ζa	Pa1	Pa2	ζa
0.00	36.3	0.0	53.4	35.0	55.5	60.7	34.3	110.6	62.4
0.05	39.9	0.0	50.5	37.8	55.7	59.8	36.9	110.7	61.9
0.10	43.9	0.0	47.3	40.8	55.9	58.9	39.5	110.9	61.3
0.15	48.5	0.0	43.8	44.1	56.2	58.0	42.3	111.1	60.8
0.20	53.9	0.0	39.9	47.5	56.5	57.0	45.1	111.3	60.2
0.25	60.2	0.0	35.4	51.2	56.9	55.9	48.1	111.5	60.0
0.30	67.9	0.0	30.4	55.2	57.3	54.8	51.3	111.8	59.1
0.35	77.1	0.0	24.7	59.5	57.9	53.7	54.6	112.1	58.5
0.40	91.3	0.0	17.7	64.1	58.5	52.6	58.0	112.5	57.9

$\delta = -15°$

k	c=0.0kN/m²			c=50kN/m²			c=100kN/m²		
	Pa1	Pa2	ζa	Pa1	Pa2	ζa	Pa1	Pa2	ζa
0.00	52.3	0.0	65.1	48.8	77.8	53.6	47.8	154.8	52.1
0.05	55.5	0.0	63.2	52.9	77.6	53.0	52.3	154.7	51.7
0.10	58.9	0.0	61.0	57.2	77.4	52.4	56.8	154.6	51.3
0.15	62.6	0.0	58.4	61.7	77.3	51.7	61.5	154.5	50.9
0.20	66.7	0.0	55.3	66.3	77.3	51.0	66.2	154.5	50.5
0.25	71.2	0.0	51.6	71.1	77.2	50.3	71.1	154.4	50.1
0.30	76.2	0.0	46.9	76.1	77.2	49.5	76.1	154.4	49.7
0.35	82.1	0.0	40.7	81.3	77.3	48.7	81.2	154.5	49.3
0.40	89.6	0.0	31.9	86.7	77.4	47.9	86.4	154.5	48.9

Table 5.1(d) Active earth pressure and angle of failure surface ($\gamma \cdot h = 100$ kN/m^2).

$\phi = 30°$, $\delta = 0°$

k	c=0.0kN/m^2			c=50kN/ m^2			c=100kN/ m^2		
	P_{a1}	P_{a2}	ζ_a	P_{a1}	P_{a2}	ζ_a	P_{a1}	P_{a2}	ζ_a
0.00	33.3	0.0	60.0	33.3	57.7	60.0	33.3	115.5	60.0
0.05	36.4	0.0	57.8	36.3	57.8	59.2	36.3	115.5	59.5
0.10	39.7	0.0	55.3	39.4	57.8	58.4	39.3	115.5	59.0
0.15	43.3	0.0	52.6	42.7	57.9	57.5	42.5	115.6	58.5
0.20	47.3	0.0	49.6	46.2	58.0	56.6	45.7	115.7	58.0
0.25	51.8	0.0	46.3	49.9	58.2	55.7	49.1	115.8	57.5
0.30	56.9	0.0	42.6	53.8	58.4	54.8	52.6	115.9	56.9
0.35	62.8	0.0	38.4	57.9	58.6	53.8	56.2	116.1	56.4
0.40	69.7	0.0	33.6	62.3	59.0	52.8	60.0	116.3	55.8

$\delta = +15°$

k	c=0.0kN/m^2			c=50kN/ m^2			c=100kN/ m^2		
	P_{a1}	P_{a2}	ζ_a	P_{a1}	P_{a2}	ζ_a	P_{a1}	P_{a2}	ζ_a
0.00	30.1	0.0	56.9	29.1	51.1	63.2	28.6	101.7	64.8
0.05	33.3	0.0	54.3	31.7	51.2	62.3	30.9	101.8	64.2
0.10	36.8	0.0	51.6	34.5	51.4	61.3	33.3	102.0	63.7
0.15	40.7	0.0	48.6	37.5	51.7	60.4	35.9	102.2	63.1
0.20	45.2	0.0	45.3	40.6	52.0	59.4	38.5	102.4	62.5
0.25	50.3	0.0	41.8	44.0	52.3	58.3	41.3	102.6	61.9
0.30	56.3	0.0	37.8	47.7	52.7	57.2	44.2	102.9	61.3
0.35	63.3	0.0	33.6	51.6	53.2	56.1	47.3	103.2	60.7
0.40	71.8	0.0	28.8	55.9	53.7	55.0	50.0	103.5	60.1

$\delta = -15°$

k	c=0.0kN/m^2			c=50kN/ m^2			c=100kN/ m^2		
	P_{a1}	P_{a2}	ζ_a	P_{a1}	P_{a2}	ζ_a	P_{a1}	P_{a2}	ζ_a
0.00	41.6	0.0	65.1	39.3	69.3	56.3	38.5	137.9	54.7
0.05	44.7	0.0	63.3	43.0	69.1	55.6	42.5	137.8	54.3
0.10	47.9	0.0	61.3	46.9	69.0	54.9	46.5	137.7	53.9
0.15	51.4	0.0	59.0	50.9	68.9	54.2	50.7	137.7	53.5
0.20	55.3	0.0	56.5	55.0	68.8	53.5	55.0	137.6	53.1
0.25	59.4	0.0	53.6	59.4	68.8	52.8	59.4	137.6	52.7
0.30	63.9	0.0	50.3	63.9	68.8	52.0	63.8	137.6	52.2
0.35	69.0	0.0	46.4	68.5	68.9	51.2	68.4	137.6	51.8
0.40	74.7	0.0	41.7	73.4	68.9	50.4	73.1	137.7	51.3

$\phi = 35°$, $\delta = 0°$

k	c=0.0kN/m^2			c=50kN/ m^2			c=100kN/ m^2		
	P_{a1}	P_{a2}	ζ_a	P_{a1}	P_{a2}	ζ_a	P_{a1}	P_{a2}	ζ_a
0.00	27.1	0.0	62.5	27.1	52.1	62.5	27.1	104.1	62.5
0.05	29.8	0.0	60.5	29.8	52.1	61.7	29.8	104.1	62.0
0.10	32.8	0.0	58.3	32.6	52.1	60.8	32.5	104.2	61.5
0.15	36.0	0.0	55.9	35.6	52.2	60.0	35.4	104.2	60.9
0.20	39.6	0.0	53.3	38.8	52.3	59.1	38.3	104.3	60.4
0.25	43.5	0.0	50.6	42.1	52.4	58.1	41.4	104.4	59.8
0.30	47.8	0.0	47.6	45.6	52.6	57.2	44.6	104.5	59.3
0.35	52.6	0.0	44.3	49.4	52.9	56.2	48.0	104.7	58.7
0.40	58.1	0.0	40.7	53.4	53.2	55.2	51.4	104.9	58.1

$\delta = +15°$

k	c=0.0kN/m^2			c=50kN/ m^2			c=100kN/ m^2		
	P_{a1}	P_{a2}	ζ_a	P_{a1}	P_{a2}	ζ_a	P_{a1}	P_{a2}	ζ_a
0.00	24.8	0.0	60.1	24.0	46.7	65.6	23.5	93.0	67.1
0.05	27.6	0.0	57.8	26.4	46.8	64.7	25.6	93.1	66.6
0.10	30.6	0.0	55.4	28.9	47.0	63.7	27.8	93.3	66.0
0.15	34.1	0.0	52.7	31.6	47.2	62.7	30.2	93.4	65.4
0.20	37.9	0.0	49.9	34.5	47.5	61.7	32.6	93.6	64.8
0.25	42.3	0.0	46.9	37.7	47.8	60.7	35.2	93.9	64.2
0.30	47.1	0.0	43.6	41.0	48.2	59.6	38.0	94.1	63.5
0.35	52.7	0.0	40.1	44.6	48.6	58.5	40.8	94.4	62.9
0.40	59.2	0.0	36.4	48.5	49.1	57.3	43.8	94.7	62.2

$\delta = -15°$

k	c=0.0kN/m^2			c=50kN/ m^2			c=100kN/ m^2		
	P_{a1}	P_{a2}	ζ_a	P_{a1}	P_{a2}	ζ_a	P_{a1}	P_{a2}	ζ_a
0.00	33.0	0.0	66.1	31.4	61.5	58.9	30.7	122.4	57.4
0.05	35.8	0.0	64.4	34.7	61.3	58.2	34.2	122.3	57.0
0.10	38.9	0.0	62.5	38.2	61.2	57.5	37.8	122.2	56.5
0.15	42.1	0.0	60.5	41.7	61.1	56.8	41.6	122.1	56.1
0.20	45.6	0.0	58.3	45.5	61.1	56.1	45.4	122.1	55.6
0.25	49.3	0.0	55.9	49.3	61.0	55.3	49.3	122.1	55.2
0.30	53.4	0.0	53.3	53.4	61.0	54.5	53.4	122.1	54.7
0.35	57.8	0.0	50.3	57.6	61.1	53.7	57.5	122.1	54.2
0.40	62.7	0.0	47.0	62.0	61.2	52.8	61.7	122.2	53.7

1. *Derivation of the earth pressure equation.* Equation (5.5) is an extension of Okabe's equation. Okabe's equation is derived by treating the ground as homogeneous and by hypothesizing that the ground fails forming a wedge. But actual ground is not homogeneous, and to take the premixing method as an example, the soil parameters vary between treated and untreated layers and between soil layers above and below the ground water level. In Equation (5.5), which has been obtained by extending Okabe's equation in order to deal with this situation, the points of change are where the constants substituted in the earth pressure and angle of failure equations are set for each layer. This extension is not as strict as Okabe's derivation.

2. *Earth pressure distribution and angle of failure surface.* To use Equation (5.5), the earth pressure strength and angle of failure surface are obtained at each layer boundary where soil properties change. In this case, the earth pressure distribution and failure line within each layer are treated as linear.

3. *Calculating the earth pressure resultant force.* The earth pressure resultant force is obtained for each layer, and it is obtained by Equation (5.6) for layer i.

$$P_i = ((p_{i-1} + p_i)/2) \, (h_i/cos \; \psi) \; \text{...} \; (5.6)$$

Where: P_{i-1}: active earth pressure of the top of layer i.

In addition, the horizontal and vertical components of the earth pressure resultant force are obtained by Equation (5.7).

$$P_{ih} = P_i \; cos \; (\psi + \delta)$$
$$P_{iv} = P_i \; sin \; (\psi + \delta) \; \text{..} \; (5.7)$$

4. *Angle of friction of the wall surface.* Normally a value $\pm \; 15°$ to $20°$ is used.

5. *Comparison with other earth pressure equations.* There are two earth pressures during an earthquake that account for both the cohesion and the angle of shear resistance. One of these is extended Okabe's earth pressure equation and the other is from Design Specifications of Highway Bridges Part V, Seismic Design (hereafter abbreviated to DSHB). Figure 5.7 shows the results of a comparison of Equation (5.5) obtained by the extended Okabe's equation and the earth pressure equation in the Design Specifications of Highway Bridges with the model in Figure 5.7. For comparison purposes, Figure 5.7 also presents the results of a calculation performed using the earth pressure equation for sandy soil, (ϕ material) and clay (c material) in the TCPHFJ. In the results of calculation of the earth pressure using the earth pressure equation for sandy soil and the earth pressure equation for clay, the earth pressure is overestimated in comparison with the earth pressure evaluated by the earth pressure equations accounting for both c and ϕ. A comparison of the results of earth pressure Equation (5.5) with those from the Design Specifications of Highway Bridges reveals that within the calculation range of the model, Equation (5.5) provides an earth pressure that is more conservative. But, because of the tendency for the earth pressure to rise as the depth increases, the earth pressure equation in the Design Specifications of Highway Bridges is more toward the conservative side than Equation (5.5) at a

greater depth. And although the results of comparisons of earth pressure Equation (5.5) with the earth pressure equation in the Design Specifications of Highway Bridges differ in this way according to conditions, it can be concluded that they provide almost identical evaluations. The above facts indicate that with this method, the earth pressure is calculated using Equation (5.5), which is an extended Okabe's earth pressure equation.

6. *Negative earth pressure*. As shown in Figure 5.7, negative earth pressure is calculated using an earth pressure equation that accounts for c – ϕ. In a range where negative earth pressure has been calculated, the earth pressure is evaluated as zero to be on the conservative side.

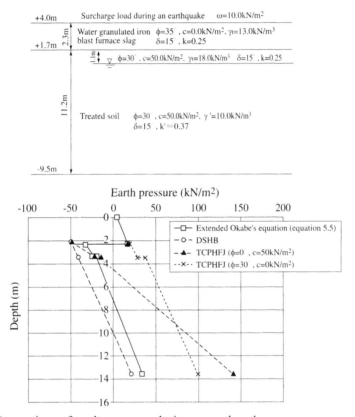

Figure 5.7 Comparison of earth pressures during an earthquake.

5.2.4. *Improvement range*

The improvement range is determined taking into account the type of the structure, external force conditions and so on. To do so, the stability of the structure under the earth pressure from the improved body, the active failure surface inside the improved body, and the overall stability including the structure and improved body must be studied.

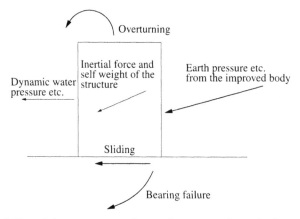

Figure 5.8 (a) Stability of the structure under earth pressure from the improved body.

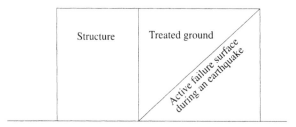

Figure 5.8 (b) Active failure surface entering the improved body.

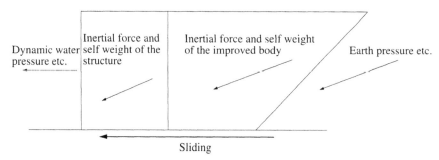

Figure 5.8 (c) Stability of the improved body including the structure.

[Commentary]
1. Because the study of the stability of the structure under earth pressure from the improved body focuses on the improvement depth, the earth pressure calculation in 5.2.3 is performed. Because this corresponds to tentatively setting or finally setting the improvement range (depth) in 5.2.2, if the study stipulated in 5.2.2 has been completed, the study in this section need not be performed.
2. The active failure surface must be completely included in the improved body as shown in Figure 5.8. This is a condition necessary to use the earth pressure equation shown in 5.2.3. And the active failure surface is that provided by Equation (5.5). When the improvement range cannot be modified for this study,

the earth pressure is calculated by the slice method proposed by Tsuchida that is presented in this section to study the stability shown in Equation (5.8). In this case the earth pressure equation in 5.2.3 is not used.

3. Sliding during an earthquake including the treated ground and the structure shown in Figure 5.9 is studied. This occurs because it is possible for the treated ground to slide as a rigid body. The standard value of the safety factor for sliding during an earthquake in this case is 1.0 or more. The coefficient of friction in a case where the ground below the treated ground is sandy soil is 0.6, and where it is clay, the cohesion of the ground under the treated ground is the standard. The water level used to study the buoyancy of the structure and the treated ground is the residual water level shown in Figure 5.10.

This study varies according to whether the ground behind the treated ground liquefies or not, and the safety factor for sliding Fs, the various external forces, and the resistance force, are provided by the following equations. In this case, the positive direction of the various external forces and resistance force are as defined in Figure 5.9.

3.1 Case where the ground behind the treated ground does not have a possibility of liquefaction (Figure 5.9 (a)).

$Fs = (R_1 + R_2 + P_{w1}) / (H_1 + H_2 + P_h + P_{w2} + P_{w3})$

$P_{w1} = (1/2) \cdot \gamma_w \cdot h_1^2$

$P_{w2} = (7/12) \cdot k \cdot \gamma_w \cdot h_1^2$

$P_{w3} = (1/2) \cdot \gamma_w \cdot h_2^2$

$H_1 = k \cdot W_1$

$H_2 = k \cdot W_2$

$P_h = (1/2) Ka \cdot \gamma' \cdot h_2^2 \cdot cos(\delta + \psi)/cos(\psi)$

$P_v = -P_h \cdot tan(\delta + \psi)$

$R_1 = f_1 \cdot W_1'$

$R_2 = f_2 \cdot (W_2' - P_v)$: (ground under the treated ground is sandy soil)

$R_2 = c \cdot l_{bc}$: (ground under the treated ground is clay)

3.2 Case there the ground behind the treated ground has a possibility of liquefaction (Figure 5.9 (b)).

$Fs = (R_1 + R_2 + P_{w1}) / (H_1 + H_2 + P_h + P_{w2})$

$P_{w1} = (1/2) \cdot \gamma_w \cdot h_1^2$

$P_{w2} = (7/12) \cdot k \cdot \gamma_w \cdot h_1^2$

$H_1 = k \cdot W_1$

$H_2 = k \cdot W_2$

$P_h = (1/2) \cdot \gamma \cdot h_2^2 + (7/12) \cdot k \cdot \gamma \cdot h_2^2$

$P_v = P_h \cdot tan \psi$

$R_1 = f_1 \cdot W_1'$

$R_2 = f_2 \cdot (W_2' + (P_v - (1/2) \cdot \gamma_w \cdot h_2^2 \cdot tan \psi))$: (ground under the treated ground is sandy soil)

$R_2 = c \cdot l_{bc}$: (ground under the treated ground is clay)

Where

γ_w: unit weight of sea water (kN/m^3)

γ':   submerged with weight of untreated soil (kN/m^3)

γ: saturated unit weight of untreated soil (kN/m^3)

k: seismic coefficient

h_1:   height of water level from the sea bottom at the front surface of the structure (m)

h_2:   height of residual water level from the sea bottom (m) (residual water level is assumed to be the ground surface to simplify the explanation)

K_a: coefficient of active earth pressure during an earthquake of untreated ground (obtained from the earth pressure equation for sandy soil)

δ: angle of friction between treated and untreated ground (*cd*) (°)

ψ: angle of the backface of treated ground to the vertical (*cd*) (positive in the counter-clockwise direction) (°)

f_1, f_2: coefficients of friction at the bottom of the structure and the bottom of the treated ground

c: cohesion of the cohesive layer beneath the treated ground (kN/m^2)

l_{bc}:   bottom length of treated ground (*bc*) (m)

In a case where the untreated ground behind the treated ground has a possibility of liquefaction, static pressure and active pressure act on the back of the treated ground from the untreated ground as shown in Figure 5.9 and sliding is studied. The static pressure is calculated by adding static water pressure to the earth pressure for a coefficient of earth pressure equal to 1.0. The dynamic pressure is calculated using Equations (5.3) and (5.4). But the unit weight of water in Equations (5.3) and (5.4) is replaced by the saturated unit weight of the soil. And because even the soil layer above the residual water level may be liquefied due to propagation of excess pore water pressure from the liquefied soil layer beneath it, it is assumed that the soil layer above the residual water level may liquefy.

The value of ψ is set based on the experience that has been set at 1:2 for the case of an existing ground behind the treated ground and at 1:4 for the case of a newly reclaimed ground behind the treated ground. Because the earth pressure equation for sandy soil does not have a solution is unobtainable in a case where the sum of ψ, δ, and the composite seismic angle θ during an earthquake is 90° or more, for convenience it is assumed that $\psi + \delta + \theta = 90°$.

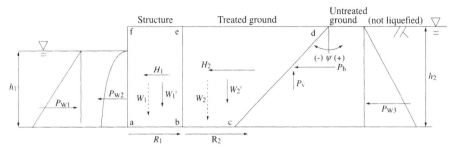

H_1:   inertia force acting on the structure (abef) (kN/m)
H_2:   inertia force acting on the treated ground (bcde) (kN/m)
P_{w1}:   resultant force of the static water pressure acting on the front surface of the structure (af) (kN/m)
P_{w2}:   resultant force of the dynamic water pressure acting on the front surface of the structure (af) (kN/m)
P_{w3}:   resultant force of the static water pressure acting on the back surface of the treated ground (cd) (kN/m)
P_h: horizontal component of the resultant forces of the active earth pressure during an earthquake from
          the untreated ground acting on the back surface of the treated ground (cd) (kN/m)
P_v: vertical component of the resultant forces of the active earth pressure during an earthquake from
          the untreated ground acting on the back surface of the treated ground (cd) (kN/m)
W_1, W_2: weight of the structure (abef) and the treated ground (bcde) (not accounting for the buoyancy generated
          by the static water pressure) (kN/m)
W_1', W_2': effective weight of the structure (abef) and the treated ground (bcde) (accounting for the buoyancy)
          (kN/m)
R_1:   friction resistance force at the bottom of the structure (ab) (kN/m)
R_2:   friction resistance force at the bottom of the treated ground (bc) (kN/m)

Figure 5.9 (a) Case where the untreated ground does not liquefy (residual water level is treated as the ground surface for convenience).

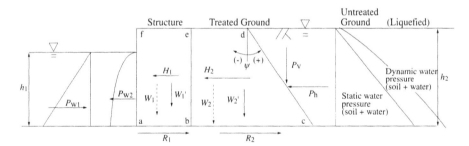

H_1:   inertia force acting on the structure (abef) (kN/m)
H_2:   inertia force acting on the treated ground (bcde) (kN/m)
P_{w1}:   resultant force of the static water pressure acting on the front surface of the structure (af) (kN/m)
P_{w2}:   resultant force of the dynamic water pressure acting on the front surface of the structure (af) (kN/m)
P_h: horizontal component of the resultant forces of the static pressure and the dynamic pressure from
          the liquefied ground acting on the back surface of the treated ground (cd) (kN/m)
P_v: vertical component of the resultant forces of the static pressure and the dynamic pressure from the
          liquefied ground acting on the back surface of the treated ground (cd) (kN/m)
W_1, W_2: weight of the structure (abef) and the treated ground (bcde) (not accounting for the buoyancy generated
          by the static water pressure) (kN/m)
W_1', W_2': effective weight of the structure (abef) and the treated ground (bcde) (accounting for the buoyancy)
          (kN/m)
R_1:   friction resistance force at the bottom of the structure (ab) (kN/m)
R_2:   friction resistance force at the bottom of the treated ground (bc) (kN/m)

Figure 5.9 (b) Case where the untreated ground liquefies.

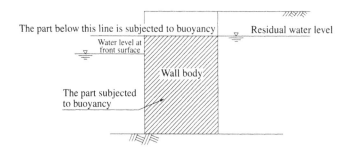

Figure 5.10 Assumption for calculating buoyancy.

4. If stability cannot be guaranteed in the above studies, the improvement range is modified or the strength of the treated sand in 5.2.2 is increased.

5. In a case where the untreated ground behind the treated ground has a possibility of liquefaction, in the form of the treated ground shown in Figure 5.11(b), hydraulic pressure from the liquefied ground acts upwards against the treated ground, reducing the effective weight of the treated ground. Consequently, it is a form that is more detrimental for sliding than the form of the treated ground in Figure 5.11(a) .

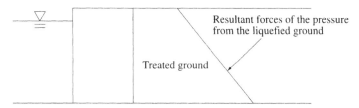

Figure 5.11(a) Force acting on the treated ground from the liquefied ground (downward).

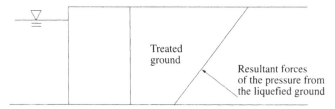

Figure 5.11(b) Force acting on the treated ground from the liquefied ground (upward).

6. In addition, in the case of a structure such as a gravity type quay wall, it is necessary to study circular. Studies other than this kind must conform to the design code for the structure concerned. But in this case, the treated ground is treated as a $c - \phi$ material.

7. If the improvement range in 2. is not adequate, the earth pressure is calculated by Tsuchida's slice method shown below, and the stability of the structure in 1., is

studied to determine the improvement range.

$$P = \Sigma(W_i \cdot k + \frac{-c \cdot l_i \sec\alpha + W_i'(\tan\alpha - \tan\phi)}{1 + \tan\alpha \cdot \tan\phi}) \quad \dots\dots\dots\dots\dots \quad (5.8)$$

Where:

P: resultant force of earth pressure per unit length (kN/m)

δ: angle of wall friction (°)

c: cohesion of the ground (kN/m²)

ϕ: angle of shear resistance (°) of the ground

α: angle of failure surface (°)

W_i: total weight of the *i*-th slice per unit length (kN/m)

W_i':   effective weight of the *i*-th slice per unit length (kN/m)

l_i: slip length of the *i*-th slice (m)

T_i: shear strength on the slip surface

N_i: normal force on the slip surface

V_i: vertical force on the right side of the slice

E_i: horizontal force on the right side of the slice

k: horizontal seismic coefficient

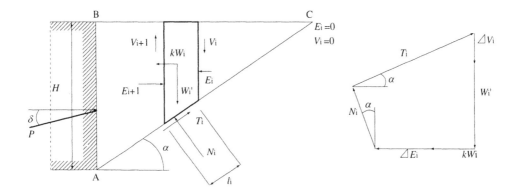

Figure 5.12 Earth pressure calculation method based on Tsuchida's method.

5.3 DESIGN FOR LIQUEFACTION PREVENTION

If the purpose is the prevention of liquefaction, the procedure shown in Figure 5.13 is followed.

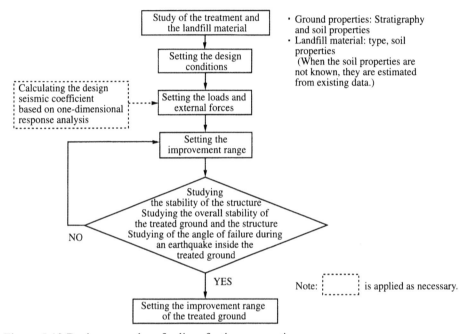

Figure 5.13 Design procedure for liquefaction prevention.

5.3.1 *External forces*
These are set in the same way as stipulated in 5.2.1.

5.3.2 *Setting the strength of the treated soil*
The design strength of the treated soil to be used for liquefaction prevention is an unconfined compressive strength of 100 kN/m^2 or more.

[Commentary]

When the purpose is liquefaction prevention, the strength of the treated soil is set so that it will be a material that does not liquefy. There is a significant relationship between the liquefaction strength and unconfined compressive strength of treated soil. It has been reported that according to the type of treated soil, liquefaction of the treated soil will not occur if its unconfined compressive strength is greater than between 50 and 100 kN/m^2. Consequently, in the case of liquefaction prevention, unconfined compressive strength is used for the criterion for the strength of treated soil. The strength of the treated soil is conservatively set at an unconfined compressive strength of 100 kN/m^2 or more. The percentage of stabilizer added at this time can be set with approximately 5% as a standard. In a case where the unconfined compressive strength of the treated soil is less than 100 kN/m², the strength that indicates tension failure must be confirmed by performing a cyclic triaxial test as shown in 3.2 (2.2) Element Test.

5.3.3 *Improvement range*
The improvement range is determined by the improvement depth and the improvement width accounting for the type of the structure and for the external force conditions. The improvement depth extends to the depth where liquefaction is predicted. The improvement width is determined accounting for the stability of the overall treated ground and the effects on the structure of liquefaction of the untreated ground.

[Commentary]
1. In the case of a structure such as a quaywall where the earth pressure is a problem, the improvement range is determined as stipulated in 5.2.4. In this case, the stability of the structure against earth pressure from the improved body is studied as shown in Figure 5.8, and the active failure surface inside the improved body is studied. The overall stability (sliding) including the structure and the improved body are also studied.
2. For structures other than the kind referred to above the design is conducted, is based on the design method for each structure. In this case, the design for each type of structure is performed by evaluating the strength of the treated ground as a c - ϕ material, and in a case where the untreated soil has a possibility of liquefaction, by also evaluating the untreated part as a liquefied material.

Specifically, as shown in Figure 5.14, the untreated part is assumed to be liquid with a unit weight that is equal to the saturated unit weight of the soil. At this time, external force is applied to the boundary with the improved body as hydraulic pressure to study its stability and set the improvement range.

Regarding the hydraulic pressure acting from the untreated part, the dynamic pressure is accounted for in the same way as in the study in 1. as shown in Figure 5.14. However, the direction of the dynamic pressure in this case is the reverse of that of the static pressure in order to conservatively estimate the stability of the foundation.

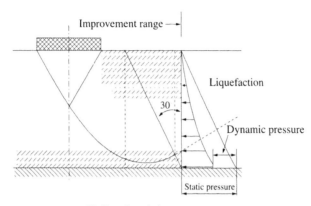

Shallow foundation

Figure 5.14 Schematic diagram of the improvement range setting.

3. When the improvement range cannot be modified in the study in 2., the strength of the treated soil is raised.

5.3.4 *Circular slip analysis*

The study of circular slip is conducted with the modified Fellenius method. The safety factor obtained from the calculation results is examined to verify that it satisfies the stipulated safety factor.

[Commentary]

1. *Circular arc method.* Slip stability analysis is generally performed based on the modified Fellenius method. In this method, the soil mass inside the slip circle is divided into several slices by vertical planes, and the forces acting on the both sides of the slice are ignored. This method is employed in TCPHFJ Part V, Chapter 6, 6.2 Stability Analysis. This method is adopted because the study can be easily performed. Another approach is to repeat the study also using the simple Bishop Method when necessary. The calculation is performed as shown in Equation (5.9) and Figure 5.15.

$$F = \frac{R\Sigma(c \cdot l + W' \cos\alpha \cdot \tan\phi)}{\Sigma W' \cdot x + \Sigma H \cdot a}$$

$$= \frac{\Sigma(c \cdot b + W' \cos^2\alpha \cdot \tan\phi) \sec\alpha}{\Sigma W \sin\alpha + \dfrac{1}{R}\Sigma H \cdot a} \quad \cdots\cdots\cdots\cdots\cdots\cdots\cdots\cdots (5.9)$$

Where:

F: safety factor against circular slip failure according to the modified Fellenius method

R: radius of the slip circle (m)

c: cohesion of soil (kN/m²)

ϕ: angle of shear resistance (°)

l: base length of a slice (m)

b: width of a slice (m)

W': effective weight of a slice per unit length (sum of soil weight and surcharge; for submerged part, use the submerged unit weight) (kN/m)

W: total weight of a slice per unit length (sum of soil weight, water weight and surcharge) (kN/m)

α : angle of the base of a slice to the horizontal (positive in the case shown in Figure 5.15) (°)

x: horizontal distance between the center of gravity of a slice and the center of the slip circle (m)

H: horizontal external force acting on the soil mass within the slip circle (hydraulic pressure, seismic force, wave pressure, etc.) (kN/m)

a: length of the arm of the horizontal external force H with respect to the center of the slip circle (m)

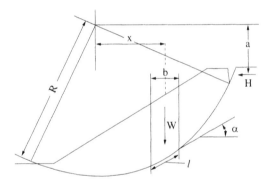

Figure 5.15 Circular slip analysis based on the modified Fellenius method.

2. *Strength parameter used for the circular slip analysis.* The strength parameters c and ϕ used to solve Equation (5.9) are the values used in 5.2.2: Strength of the treated soil and in 5.5: Laboratory mix proportion text.

5.4 MIX PROPORTION DESIGN

The mix proportion design of the treated soil is established by performing an appropriate laboratory mix proportion test. It is considered in the mix proportion design that the field strength will be lower than the laboratory strength.

[Commentary]
The mix proportion design is performed as shown in Figure 5.16.

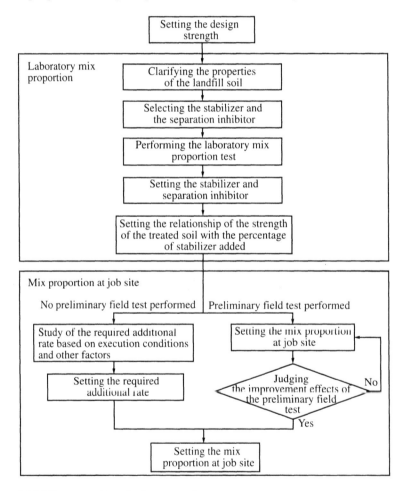

Figure 5.16 Mix proportion design procedure.

The many factors influencing the improvement effects of the premixing method include the properties of the landfill soil (natural water content, pH, gradation), the type and quantity of the stabilizer used, degree of mixing, and curing conditions. Because of the complexity of their effects, it is difficult to directly obtain field strength after reclamation by laboratory test. But it is possible to estimate the post-reclamation field strength by performing an appropriate laboratory mix proportion test accounting for all factors that cause improvement effects and the results of past executions. Consequently, the mix proportion design of treated soil involves designing the mix proportion at the job site based on the results of a laboratory mix proportion test carried out to decide the type of cement used and the percentage of the cement added that will satisfy the design strength. And where necessary, a preliminary field test is performed to complete the design of the mix proportion.

1. *Laboratory mix proportion test.* A laboratory mix proportion test is done to obtain the relationship of the strength of the treated soil with the percentage of stabilizer added, in order to determine the percentage of stabilizer added that will provide the treated soil with the required strength. The above relationship is greatly influenced by the soil type, density, and other test conditions. Therefore, the test conditions for the mix proportion test must be set carefully so that they correspond to actual conditions at the site.

1.1 The type of stabilizer and the type of separation inhibitor are set based on the results of the test because they influence the strength of the treated soil.

1.2 The treated ground is designed by performing a laboratory test suited to its purpose in order to obtain the relationship of the strength with the percentage of stabilizer added.

- In a liquefaction prevention case, an unconfined compression test is performed to obtain the relationship of the unconfined compressive strength with the percentage of stabilizer added.

- In an earth pressure reduction case, a consolidated drained triaxial test is performed to obtain the relationship of cohesion c and angle of shear resistance ϕ with the percentage of stabilizer added.

2. *Setting the mix proportion at job site.* Preliminary field test is performed as necessary to clarify the scattering of the density and strength of the treated ground after reclamation and differences between field strength and laboratory strength.

2.1 The strength obtained from a laboratory mix proportion test of treated soil provides an index of strength that can be achieved under certain test conditions; it does not directly provide the strength that can be achieved on site. For this reason, the mix proportion at a job site is set using a required additional rate that corrects for differences with on-site execution conditions.

2.2 The clarification of the differences between field strength and laboratory strength is important when setting the required additional rate for the mix proportion at the job site. Based on past experience, the laboratory strength is greater than the field strength and the required additional rate applied $\alpha = 1.1$ to 2.2. So the required additional rate α is the ratio of the laboratory strength and the field strength, and is defined as the ratio of the unconfined compressive strength.

5.5 LABORATORY MIX PROPORTION TEST

The laboratory mix proportion test is done to obtain the relationship of the type and percentage of the stabilizer added and the type and quantity of the separation inhibitor added with the strength of the treated soil.

[Commentary]
1. *Selection of the stabilizer used.* The stabilizer is used to increase the cohesion of the landfill soil, which is dependent upon the chemical reaction between the landfill soil and the stabilizer. For this reason, the selected stabilizer must be a material that is effective and efficient when used with the landfill soil. The types of stabilizer available include ordinary portland cement, blast furnace slag cement, and special cements made for use in soil improvement. Blast furnace slag cement type B has been used in many past executions of the premixing method.
2. *Selection of the separation inhibitor used.* When mixed soil made by adding stabilizer to landfill soil is placed under water without a separation inhibitor, the soil and the stabilizer may separate. Separation inhibitors that are used to prevent the separation of the landfill soil and the stabilizer include cellulose and acrylic types, and soluble high polymer materials. In past executions of the premixing method, polyacrylamide has been used frequently. As the proportion by mass of dry soil, it has been added at a rate of 40 mg/kg when dissolved in fresh water and at a rate of 90 mg/kg when dissolved in sea water.
3. *Selection of the landfill soil.* In many past cases, the soil used has had a natural water content of 15% or less and a fines content of 15% or less. But in some cases, landfill soil with a high water content (20% or more) has been used.
4. *Laboratory test.* As an index of improvement effects, an unconfined compression test has been conducted for liquefaction prevention applications and consolidated drained triaxial test for earth pressure reduction applications. The strength properties of treated soil are governed by the percentage of the stabilizer added and by the density of the landfill soil. Therefore, the test is performed on specimens prepared by varying the percentage of cement added and their density. The density of the treated soil for the test must be estimated accounting for past executions.
5. *Unconfined compressive strength – c and ϕ relationship.* Figure 5.17 shows the Mohr's stress circle obtained from the triaxial compression test of the specimens before and after treatment. Since the cohesion of the treated soil is increased from that of the untreated soil without changing its original angle of shear resistance, the shear strength of the treated soil can be represented by the following equation.

$$\tau_d = c_d + \sigma' \, tan\phi_d \quad\text{(5.10)}$$

If the maximum principal stress difference of the drained triaxial test with no confining pressure is represented as q_{cd}, the value of c_d in the first term in

Equation (5.10) can be represented by Equation (5.11) using q_{cd} and the angle of shear resistance (ϕ_d).

$$c_d = \frac{q_{cd}}{2} \cdot \frac{1 - sin\phi_d}{cos\phi_d}$$

$$= \frac{q_{cd}}{2tan(45° + \phi_d/2)} \quad\text{...} (5.11)$$

Where it is assumed that $q_{cd} \fallingdotseq q_{ud}$ and the relationship between the angle of shear resistance of the treated soil (ϕ_d) and the angle of shear resistance of the untreated soil (ϕ_{d0}) is represented as $\phi_d \fallingdotseq \phi_{d0}$, ϕ_d is replaced by ϕ_{d0} in Equation (5.11), and Equation (5.11) is represented by Equation (5.12):

$$c_d = \frac{q_{ud}}{2tan(45° + \phi_{d0}/2)} \quad\text{...} (5.12)$$

Where ϕ_{d0} is the value obtained by triaxial test.

Consequently, it is possible to estimate the shear strength of the treated soil τ_d using Equation (5.13) based on Equation (5.12). In this case the angle of shear resistance of the untreated soil (ϕ_{d0}) and the unconfined compressive strength of the treated soil (q_{ud}) must be known in advance.

$$\tau_d = \frac{q_{ud}}{2tan(45° + \phi_{d0}/2)} + \sigma' tan\phi_{d0} \quad\text{..} (5.13)$$

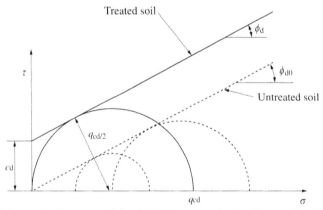

Figure 5.17 Schematic diagram of the Mohr's stress circle of treated soil and untreated soil specimens.

Figure 5.18 shows the results of a comparison of the estimated and the measured cohesion (c_d) done in order to study the appropriateness of Equation (5.13). This figure shows that within a range of dry density 1.2 g/cm$^3 \leqq \rho_d <$ 1.7 g/cm^3, those values are almost identical, showing that it is possible to roughly estimate the shear strength of the treated soil based on Equation (5.13).

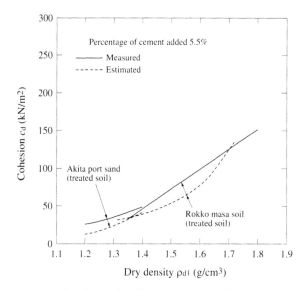

Figure 5.18 Comparison of estimated and measured cohesion c_d.

5.6 MIX PROPORTION AT JOB SITE

The mix proportion at a job site, which is the mix proportion used for the execution, must be determined by correcting the difference between the laboratory strength and the field strength.

[Commentary]

The mix proportion at the job site is the mix proportion used for the execution of reclamation. First, strength is obtained by multiplying the design strength by the required additional rate (α) used to correct the difference between the laboratory strength and the field strength. Then the percentage of stabilizer added is determined based on the unconfined compressive strength-percentage of stabilizer added relationship shown in Figure 5.19 obtained from laboratory test. The percentage of stabilizer added that is obtained at that stage is used to prepare the landfill material at the site.

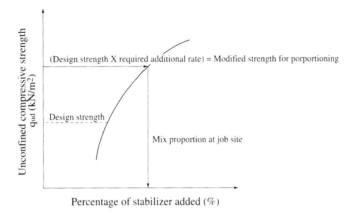

Figure 5.19 Mix proportion at job site.

It is desirable that a preliminary field test should be performed to obtain an accurate required additional rate to determine the mix proportion at the job site. However, because it is not always possible to perform a preliminary field test, a method of setting the job-site mix proportion without performing a preliminary field test (particularly, obtaining the required additional rate (α)) is explained below.

1. The design strength is obtained by converting the cohesion set by the design of the treated ground to the unconfined compressive strength using Equation (5.12).

2. The laboratory strength is the mean unconfined compressive strength obtained by laboratory test.

3. The field strength is the mean unconfined compressive strength at the site obtained by a mix proportion equal to that of the laboratory test. But the field strength is smaller than the laboratory strength, because of differences in the degree of mixing of the materials, the separation of the stabilizer in the water, and differences in execution methods and curing environments.

4. The required additional rate (α) is used to correct for the difference between the laboratory strength and the field strength, and it is defined by Equation (5.14).

$$\alpha = \overline{q_{udl}}/\overline{q_{udf}} \quad (5.14)$$

α : required additional rate

$\overline{q_{udl}}$: laboratory strength (kN/m^2)

$\overline{q_{udf}}$: field strength (kN/m^2)

Table 5.2 shows the required additional rate (α) obtained from past preliminary field test and executions. According to the executions listed in the table, the required additional rate ranges from 1.1 to 2.2. The required additional rate is set in three stages, as explained below. Table 5.3 presents required additional rates used for other cement stabilization reclamation methods for reference. Factors that influence the required additional rate include purpose of the improvement, reclamation method, water depth, mixing method, design strength, and soil properties (including the fines content).

Table 5.2 Examples of required additional rates.

Year	Name of Execution	Type of Landfill Soil	Stabilizer Mixing Method	Reclamation Execution Method	Objective	Percentage of Stabilizer Added (%)	Quantity of Separation inhibitor Added mg/kg	Design Strength q_{ud} kN/m²	Required Additional Rate α
1987	1/7 direct placement method (preliminary)	Rokko masa soil	Belt conveyor mixing	1/7 scale open bottom sand barge		5.5	50	100	2.2
1988	Rokko island reclamation (preliminary)	Tokushima gravel	Back hoe mixing	Pushed into place with a bulldozer on dry land		4	0	100	2.2
1989	1/36 direct placement method (preliminary)	Sengenyama pit sand	Belt conveyor mixing	1/36 scale open bottom sand barge		8	200	400	1.1
1990	Steel sheet pile cell installation (preliminary)	Sengenyama pit sand	Belt conveyor mixing	Chute reclamation at a depth of 10 m		7.3	30	400	1.1
1992 to 1993	Kisarazu Artificial Island on Tokyo Bay Aqualine (TTB)	Sengenyama pit sand	Belt conveyor mixing	Chute reclamation at a depth of 2.5 m	Liquefaction prevention Bearing capasity	7.5	90	400	*) (1.0)
1993	Niigata Airport extension	Soil dredged from Niigata Harbor	Belt conveyor mixing	Pushed in place by bulldozers on dry land	Liquefaction prevention	3	75	100	1.7
1993	Port of Ishikari	Soil dredged from Ishikari Harbor	Belt conveyor mixing	Pushed in place by bulldozers on dry land	Liquefaction prevention	4	75	50	2.0
1995 to 1996	Rokko Island – 10 m quaywall	Masa soil	2-shaft mixer + belt conveyor mixing	Deep parts: open bottom bucket Shallow parts: pushed by bulldozers	Earth pressure reduction	7.5	75	100	1.8
1996	Rokko Island Ferry Jetty	Masa soil	Belt conveyor mixing	Deep part: clam shell Shallow part: pushed by dry land bulldozer + long arm back hoe	Earth pressure reduction	7.5	75	100	1.8

*) The required additional rate was not necessary because the preliminary field test was conducted before the execution.

Table 5.3 Stabilizer types, mixing methods and required additional rates for cement stabilization methods.

(Modified from source document)

State of the stabilizer added	Type of soil improved	Execution machinery	Required Additional Rate α
Powder	Soft soil	Stabilizer	1.25~2.00
		Back hoe	1.43~3.33
	Sludge	Clam shell	2.00~5.00
	High water content organic soil	Back hoe	
Slurry	Soft ground	Stabilizer	1.25~2.00
		Back hoe	1.43~2.50
	Sludge	Vessel	1.25~2.00
		Working vehicle for mad	1.43~3.33
	High water content organic soil	Clam shell and back hoe	1.67~3.33

The required additional rates determined with reference to these factors and to Table 5.2 are shown in Figure 5.20. The required additional rate is determined according to following groups of execution conditions.

Group [1]: Execution conditions allowing homogeneous ground to be formed easily

Group [2]: Ordinary execution conditions

Group [3]: Execution conditions lowing the homogeneity of the treated ground in comparison with group [1] and group [2].

The execution condition is classified to one of the above groups with the ranks A, B, and C. The ranks correspond to the influence of each factor in Table 5.4. The rank A corresponds to the high possibility of homogeneous reclamation. On the other hand, the rank C corresponds to the low possibility of the homogeneous reclamation. When all the factors correspond to the rank A in Table 5.4, the execution site is classified to the group [1]. And the required additional rate is determined in the range of ① in Figure 5.20. When at least one factor corresponds to the rank C, the execution site is classified to the group [3]. And the required additional rate is determined in the range of ③ in Figure 5.20. In other cases, the execution site is classified to the group [2]. And the required additional rate is determined in the range of ② in Figure 5.20.

Table 5.4 Deciding factors for required additional rate (reference).

Factor	Judgement		
	A ~	B ~	C
Purpose of Improvement	Liquefaction prevention		Earth pressure reduction
Mixing method	Mixer	Belt conveyor	Back hoe
Water depth	Shallower than 5 m	Shallower than 10 m	Deeper than 10 m
Reclamation method	Chute	Bulldozer	Direct placement etc.
Design strength	Min. 400 kN/m^2	Min. 100 kN/m^2	Max. 100 kN/m^2
Fines content *)	Large	Medium	Small

All factors are judged as A   Use group [1]   Range ① of Fig. 5.20.
At least one factor is judged as C   Use group [3]   Range ② of Fig. 5.20.
Others   Use group [2]   Range ③ of Fig. 5.20.
*)   Basically, sandy soil with a fines content of 15% or less

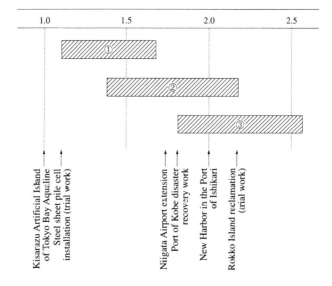

Figure 5.20 Required additional rates.

Although the required additional rate can be determined with the above procedure, it is desirable that the rate be determined with the preliminary field test. In all the past execution of the premix method, the required additional rate was determined with the preliminary field test.

5.7 EXAMPLES OF DESIGN

5.7.1 *Design where the ground behind the treated ground does not have a possibility of liquefaction*
1. *Items studied.* The study is performed using the section shown in Figure 5.21. The following items are studied.
1.1 Wall body stability
1.2 Overall stability (when the ground behind the treated ground does not have a possibility of liquefaction)

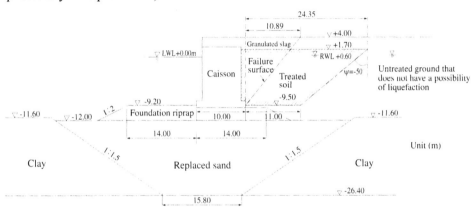

Figure 5.21 Section diagram.

2. *Design conditions*
2.1 Structure	Caisson type quaywall	
2.2 Water depth	Design water depth	-9.50 m
2.3 Levels	Caisson bottom	- 9.50 m
	Caisson crown height	+ 2.00 m
	Coping crown height	+ 4.00 m
	Backfill surface	+ 4.00 m
2.4 Surcharge	During an earthquake	10 kN/m^2
2.5 Design seismic coefficient	Horizontal seismic coefficient	$k_h = 0.25$
2.6 Unit weight	Reinforced concrete	$\gamma_c = 24.0$ kN/m^3
	Unreinforced concrete	$\gamma_c = 22.6$ kN/m^3
	Sea water	$\gamma_w = 10.1$ kN/m^3

2.7 Soil properties   treated ground
        +4.00 m to +1.70 m   Granulated slag
        $\phi = 35°$, $\gamma_t = 13$ kN/m^3, $\gamma = 17$ kN/m^3.
        +1.70 m to -9.50 m   Premixed soil
        $\phi = 30°$, $\gamma_t = 18$ kN/m^3, $\gamma = 20$ kN/m^3.
        $c = 50$ kN/m^2
        Ground behind the treated ground
        +4.00 m to +0.60 m $\phi = 30°$, $\gamma_t = 18$ kN/m^3.
        +0.60 m to -9.50 m $\phi = 30°$, $\gamma = 20$ kN/m^3.

2.8 Coefficient of friction

Caisson $\mu = 0.7$ (friction increasing mat and riprap)
Improved ground: $\mu = 0.6$

2.9 Others   Other design conditions follow with the TCPHFJ

Table 5.5 Weight and moment of the caisson.

	w(kN/m)	w ·X(kN ·m/m)	w ·y(kN ·m/m)
Total caisson weight	2505.1	13532.0	16456.3
Buoyancy acting on the caisson	925.2	5053.4	
Effective weight of the caisson	1579.9	8478.6	

3. Stability of wall body

3.1 *Active earth pressure.* The active earth pressure is calculated using Equation (5.5) from 5.2.3: Earth Pressure. In the section in the example, $\psi = 0°$ and $\beta = 0°$ while other conditions are calculated under the conditions shown in Figure 5.22. The seismic coefficient is, above the residual water level, the seismic coefficient in the air ($k_h = 0.25$), and below the residual water level, the apparent seismic coefficient ($k_h = 0.37$). Table 5.6 shows the active earth pressures and the horizontal earth pressures that are their horizontal component at the depths needed for the calculation. In addition, Figure 5.22 shows the distribution of the horizontal earth pressure and the residual water pressure that act on the wall body.

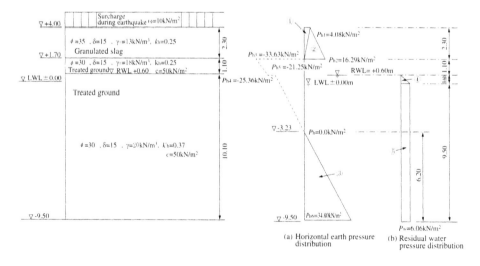

(a) Horizontal earth pressure distribution

(b) Residual water pressure distribution

Figure 5.22 Horizontal earth pressure distribution and residual water pressure distribution.

Table 5.6 Active earth pressure and horizontal earth pressure calculation table.

	Eleva-tion	A	B	C	Angle of Failure ζ (°)	μ	Active Earth Pressure p_{ai} (kN/m^2)	Horizontal Earth Pressure $p_{ai}\,cos\,(\psi_2 + \delta)$ (kN/m^2)
p_{h1}	+4.00	0.485	1.089	0.093	46.87	31.26	4.23	4.08
p_{h2}	+1.70	0.485	1.089	0.093	46.87	31.26	16.86	16.28
p_{h3}	+1.70	0.485	4.204	-0.771	61.94	-3.87	-34.82	-33.63
p_{h4}	+0.60	0.485	3.280	-0.524	60.33	-0.67	-26.26	-25.36
p_{h5}	+0.60	0.578	3.077	-0.401	58.35	3.30	-22.00	-21.25
p_{h6}	-9.50	0.578	1.745	-0.045	51.06	17.87	36.03	34.80

3.2 Horizontal force and moment generated by the active earth pressure and the residual water pressure

Table 5.7 is the calculation table for the horizontal forces and the moment generated by the active earth pressure and residual water pressure.

Table 5.7 Active earth pressure and residual water pressure generated lateral force and moment calculation table.

	Equation	H (kN/m)	y (m)	H·y (kN·m/m)
①	$1/2 \times 4.08$kN/m$^2 \times 2.30$m	4.7	12.73	59.8
②	$1/2 \times 16.29$kN/ m$^2 \times 2.30$m	18.7	11.97	223.8
③	$1/2 \times 34.80$kN/ m$^2 \times 6.27$m	109.1	2.09	228.0
	Sub-totals	132.5		511.6
④	$1/2 \times 6.06$kN/ m$^2 \times 0.60$m	1.8	9.70	17.5
⑤	6.06kN/ m$^2 \times 9.50$m	57.6	4.75	273.6
	Totals	191.9		802.7

3.3 *Vertical force and moment generated by earth pressure*
3.3.1 Vertical force
$$P_v = P_h \times tan\delta$$
$$= 132.5 \times 0.268 = 35.5\text{kN/m}$$
3.3.2 Moment
$$M_v = P_v \times x$$
$$= 35.5 \times 10.00 = 355.0\text{kN·m/m}$$
4. *Dynamic water pressure acting on the front surface of the wall body*
4.1 Resultant force of dynamic water pressure acting on the front surface of the wall
$$P_{dw} = 7/12 \; k \cdot \gamma_w \cdot H^2$$
$$= 7/12 \times 0.25 \times 10.1 \times 9.5^2 = 132.9\text{kN/m}$$

4.2 Acting point of the resultant force of dynamic water pressure (distance from the bottom of the front toe of the wall body)

$$h_{dw} = 2/5 \cdot H$$
$$= 2/5 \times 9.5$$
$$= 3.8m$$

4.3 Moment of dynamic water pressure acting on the front surface of the wall body

$$M_{dw} = 132.9 \times 3.8 = 505.0kN \cdot m/m$$

5. *Stability study*

5.1 Stability against sliding

ΣW= effective weight of caisson + vertical force generated by earth pressure
$$= 1579.9 + 35.5$$
$$= 1615.4kN/m$$
$$\mu = 0.7$$

ΣH= total weight of caisson $\times$ design seismic force + earth pressure, horizontal force generated by residual water pressure + dynamic water pressure acting on the front surface of the wall
$$= 2505.1 \times 0.25 + 191.9 + 132.9$$
$$= 951.1kN/m$$

$$F_s = \frac{\Sigma Wk \times \mu}{\Sigma H} = \frac{1615.4 \times 0.7}{951.1} = 1.19 > F_{sa} = 1.0$$

5.2 *Stability against overturning*

ΣM_r = Moment acting on the caisson + moment of the vertical force generated by the earth pressure
$$= 8478.6 + 355.0$$
$$= 8833.6kN \cdot m/m$$

ΣM_a = moment of the (total weight of caisson $\times$ design seismic coefficient) +moment of the earth pressure and horizontal force generated by the residual water pressure + moment of the dynamic water pressure acting on the front surface of the wall body
$$= 16456.3 \times 0.25 + 802.7 + 505.0$$
$$= 5421.8kNm/m$$

$$F_a = \frac{\Sigma M_r}{\Sigma M_a} = \frac{8833.6}{5421.8} = 1.63 > F_{sa} = 1.1$$

5.3 *Bottom reaction*

$$x = \frac{\Sigma M_r - \Sigma M_a}{\Sigma W} = \frac{8833.6 - 5421.8}{1615.4} = 2.112m$$

$$e = \frac{B}{2} - x = \frac{10.00}{2} - 2.112 = 2.888m > \frac{B}{6} = 1.667m$$

$$P_{max} = \frac{2}{3} \cdot \frac{\Sigma W}{x} = \frac{2}{3} \times \frac{1615.4}{2.112} = 509.9 \text{kN/m}^2$$

$b' = 3 x = 3 \times 2.112 = 6.336\text{m}$

6. *Bearing capacity against eccentric inclined load*
   From the results of the stability study:
   $B = 10.0\text{m}, x = 2.112\text{m}, b' = 6.336\text{m}, P_{max} = 509.9\text{kN/m}^2$
   $\Sigma H = 947.8\text{kN/m}$
   Therefore:

   $2 x = 4.224\text{m}$

   $$q = \frac{P_{max} \cdot b'}{4 x} = \frac{509.9 \times 6.336}{4 \times 2.112} = 382.4 \text{kN/m}^2$$

Figure 5.23 shows the results of studying the bearing capacity against eccentric inclined load during an earthquake based on the Bishop Method under the above conditions.

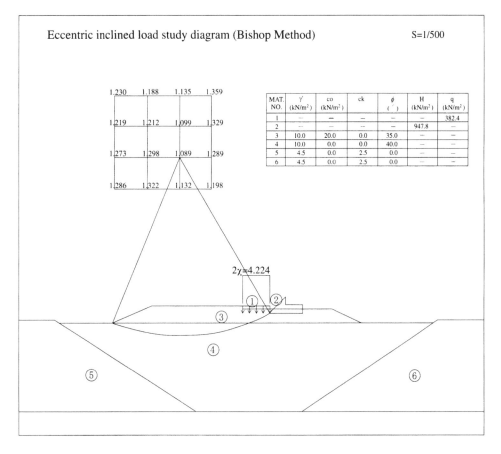

Figure 5.23 Study of the bearing capacity against eccentric inclined load.

7. *Study of overall stability*
7.1 *Calculation of the volume of the treated ground*

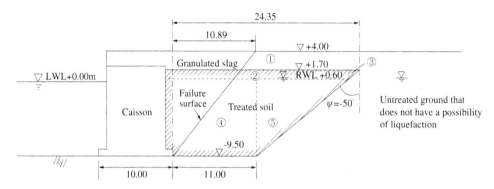

Figure 5.24 Volume of the treated ground.

Table 5.8 Treated ground volume calculation table.

	Equation	V (m³/m)
① V_1	$24.35 \times 2.3 \times 1.0$	56.00
② V_2	$23.04 \times 1.1 \times 1.0$	25.34
③ V_3	$1.31 \times 1.1 \div 2 \times 1.0$	0.72
④ V_4	$11.00 \times 10.1 \times 1.0$	111.10
⑤ V_5	$12.04 \times 10.1 \div 2 \times 1.0$	60.80

7.2 *Total weight of the treated ground*

Table 5.9 Treated ground weight calculation table.

	Equation	w (kN/m)
①	$56.00 \text{m}^3/\text{m} \times 13.0 \text{kN/m}^3$	728.0
②	$25.34 \text{m}^3/\text{m} \times 18.0 \text{kN/m}^3$	456.1
③	$0.72 \text{m}^3/\text{m} \times 18.0 \text{kN/m}^3$	13.0
④	$111.10 \text{m}^3/\text{m} \times 20.0 \text{kN/m}^3$	2222.0
⑤	$60.80 \text{m}^3/\text{m} \times 20.0 \text{kN/m}^3$	1216.0
	Totals	4635.1

7.3 *Buoyancy acting on the treated ground*
7.3.1 Volume subjected to the buoyancy acting on the treated ground
$$V = ④ + ⑤ = 111.10 + 60.80 = 171.90 \text{ m}^3/\text{m}$$
7.3.2 Buoyancy acting on the treated ground
$$F = 171.90 \text{ m}^3/\text{m} \times 10.1 \text{kN/m}^3 = 1736.2 \text{kN/m}$$

7.4 *Effective weight of the treated ground*

$$W_2' = 4635.1 - 1736.2 = 2898.9 \text{kN/m}$$

7.5 *Earth pressure during an earthquake acting on the treated ground from the untreated ground*

7.5.1 *Active earth pressure.* The active earth pressure is calculated assuming that $\psi = -50°$, and $\beta = 0°$, and the seismic coefficient and other conditions shown in Figure 5.24. Table 5.10 shows the active earth pressure at depths necessary for the calculation, and Figure 5.25 shows its distribution.

Table 5.10 Active earth pressure calculation table.

	Elevation	Coefficient of Active Earth Pressure Ka	Angle of Failure $\zeta(°)$	$\gamma \cdot h + w$ (kN/m^2)	Active Earth Pressure $p_{ai} = Ka\,(\gamma \cdot h + \omega)\cos\psi$ (kN/ m^2)
p_{h1}	+4.00	0.180	26.25	10.0	1.16
p_{h2}	+0.60	0.180	26.25	71.2	8.24
p_{h3}	+0.60	0.299	21.93	71.2	13.67
p_{h4}	-9.50	0.299	21.93	172.2	33.07

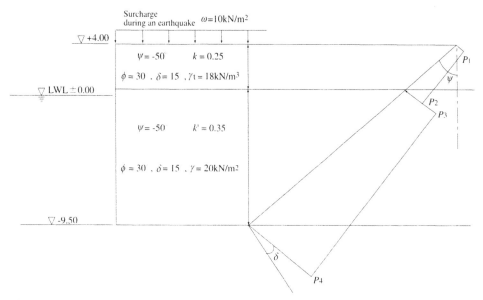

Figure 5.25 Distribution of earth pressure from the untreated ground to the treated ground.

7.5.2 Horizontal and vertical components of force of the earth pressure. The horizontal and vertical components of force of the earth pressure are as shown in Table 5.11.

Table 5.11 Horizontal and vertical components of force of the earth pressure calculation table.

Elevation	Active Earth Pressure	Elevation of the Acting Point of the Earth Pressure	Resultant Force of Earth Pressure	Horizontal Component of Force	Vertical Component of Force
	p_1	$h_1/cos\psi$	p_1	$p_h = p_1 \, cos(\delta+\psi)$	$p_v = p_1 \, sin(\delta+\psi)$
+4.00	1.16				
		5.29	24.9	20.4	-14.3
+0.60	8.24				
+0.60	13.67				
		15.71	367.1	300.7	-210.6
-9.50	33.07				
			Totals	$321.1 kN/m^2$	$-224.9 kN/m^2$

7.6 *Water pressure*

7.6.1 Water pressure acting on the front surface of the wall body
$$P_{w1} = (1/2) \, \gamma_w \, h_1^2 = (1/2) \times 10.1 \times 9.5^2 = 455.8 kN/m$$

7.6.2 Resultant force of the dynamic water pressure acting on the front surface of the wall body
$$P_{w2} = 7/12 \, k_h \, \gamma_w \, h_1^2$$
$$= 7/12 \times 0.25 \times 10.1 \times 9.5^2$$
$$= 132.9 kN/m$$

7.6.3 Water pressure acting on the back surface of the treated ground
$$P_{w3} = (1/2) \, \gamma_w \, h_2^2 = (1/2) \times 10.1 \times 10.1^2 = 515.2 kN/m$$

7.7 *Stability against sliding*

W_1' = effective weight of the caisson = 1579.9 kN/m

μ_1 = coefficient of friction between the caisson and the foundation = 0.7

$R_1 = W_1' \, \mu_1 = 1579.9 \times 0.7 = 1105.9 kN/m$

$W_2' + P_V$ = effective weight of the treated ground + vertical component of the earth pressure
= 2898.9 - 224.9 = 2674.0 kN/m

μ_2 = coefficient of friction between the treated ground and the foundation = 0.6

$R_2 = W_2' \, \mu_2 = 2674.0 \times 0.6 = 1604.4 kN/m$

$\Sigma H = (H_1 + H_2)$ = (total weight of the caisson + total weight of the treated ground) $\times$ design seismic coefficient
= (2505.1 + 4635.1) $\times$ 0.25 = 1785.1 kN/m

$$F_a = \frac{R_1 + R_2 + P_{w1}}{\Sigma H + P_h + P_{w2} + P_{w3}}$$

$$= \frac{1105.9 + 1604.4 + 455.8}{1785.1 + 321.1 + 132.9 + 515.2} = 1.15 > F\mathrm{sa} = 1.0$$

5.7.2 Design where the ground behind the treated ground has a possibility of liquefaction

1. *Items studied.* The study is performed using the section shown in Figure 5.26. In a case where the ground behind the treated ground has a possibility of liquefaction, the water level must be set at the ground surface on the right side of point B in Figure 5.27 even if the residual water level is deeper than the ground surface. This is necessary because liquefaction may be caused by propagation of excess pore water pressure even in ground shallower than the residual water level. On the left side of point B, water level is set at the residual water level.

In the design section studied, the effects of the rise in the water level caused by liquefaction of the untreated ground do not extend to the wall body as shown in Figure 5.26. The wall body stability results are, therefore, identical to those in the previous Subsection 5.7.1. Consequently, the study on the wall body stability is omitted and the report continues with the overall stability study.

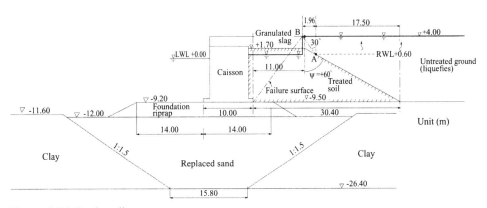

Figure 5.26 Section diagram.

2. Overall stability
2.1 Volume of the treated ground

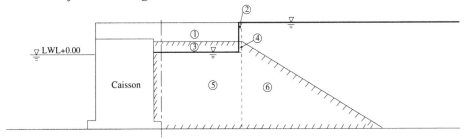

Figure 5.27 Volume of the treated ground.

Table 5.11 Treated ground volume calculation table.

	Equation	V (m^3/m)
① V_1	$10.94 \times 2.3 \times 1.0$	25.16
② V_2	$0.06 \times 2.3 \times 1.0$	0.14
③ V_3	$10.94 \times 1.1 \times 1.0$	12.03
④ V_4	$0.06 \times 1.1 \times 1.0$	0.07
⑤ V_5	$19.40 \times 11.2 \div 2 \times 1.0$	108.64
⑥ V_6	$11.00 \times 10.1 \times 1.0$	111.10

2.2 Total weight of the treated ground

Table 5.12 Treated ground weight calculation table.

	Equation	W (kN/m)
①	$25.16\text{m}^3/\text{m} \times 13\text{kN/ m}^3$	327.1
②	$0.14\text{m}^3/\text{m} \times 17\text{kN/ m}^3$	2.4
③	$12.03\text{m}^3/\text{m} \times 18\text{kN/ m}^3$	216.5
④	$0.07\text{m}^3/\text{m} \times 20\text{kN/ m}^3$	1.4
⑤	$108.64\text{m}^3/\text{m} \times 20\text{kN/ m}^3$	2172.8
⑥	$111.10\text{m}^3/\text{m} \times 20\text{kN/ m}^3$	2222.0
	Totals	4942.2

2.3 Buoyancy acting on the treated ground
2.3.1 Volume subjected to the buoyancy acting on the treated ground
$V = ② + ④ + ⑤ + ⑥ = 0.14 + 0.07 + 108.64 + 111.10 = 219.95$ m^3/m
2.3.2 Buoyancy acting on the treated ground
$F = 219.95$ m^3/m $\times 10.1$kN/m$^3 = 2221.50$kN/m
2.4 Effective weight of the treated ground
$W_2' = 4942.20 - 2221.50 = 2720.70$kN/m
2.5 Horizontal and vertical components of force acting on the treated ground from liquefied ground
2.5.1 Horizontal component of force acting on the treated ground from liquefied ground. The horizontal component of force acting on the treated ground from liquefied ground is categorized as static pressure and as dynamic pressure. Static pressure is calculated by adding the static water pressure to the earth pressure with a coefficient of earth pressure of 1.0. The active pressure is calculated by replacing the unit weight of water with the saturated unit weight of the soil.

Table 5.13 is the horizontal component of force calculation table. Figure 5.28 presents the distribution of the horizontal component of force acting on the treated ground from liquefied ground.

Table 5.13 Horizontal component of force calculation table.

Elevation	Static Pressure p_i	Height of Action h_2	Resultant Force of Static Pressure	Resultant Force of Dynamic Pressure $7/12k_h\gamma h_2^2$
+4.00	10.0			
-9.50	280.0	13.5	1957.5	531.6
		Totals		P_h=2489.1kN/m^2

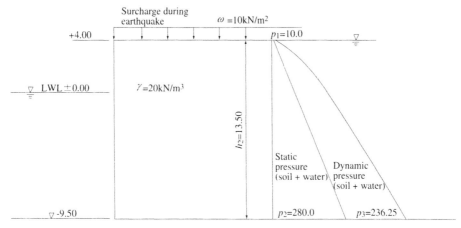

Figure 5.28 Distribution of horizontal component of force acting on treated ground from liquefied ground.

2.5.2 *Vertical component acting on the treated ground from liquefied ground*
   $P_v = P_h \tan\psi = 2489.1 \times \tan 60° = 4311.2$kN/m^2
2.6 *Water pressure*
2.6.1 *Water pressure acting on the front surface of wall body*
   $P_{w1} = (1/2) \gamma_w h_1^2 = (1/2) \times 10.1 \times 9.5^2 = 455.8$kN/m
2.6.2 *Resultant force of dynamic water pressure acting on the front surface of the wall body*
   $P_{w2} = 7/12 k_h \gamma_w h_1^2 = 7/12 \times 0.25 \times 10.1 \times 9.5^2 = 132.9$kN/m
2.7 *Stability against sliding*
   W_1' = effective weight of caisson = 1579.9 kN/m
   μ_1 = coefficient of friction between the caisson and the foundation = 0.7
   $R_1 = W_1' \mu_1 = 1579.9 \times 0.7 = 1105.9$kN/m
   W_2' + P_v = effective weight of treated ground + P_v-(1/2) $\gamma_w h_2^2 \tan\psi$
      = 2720.7 + 4311.2 – (1/2)$\times$(10.1)$\times$(13.5)$^2 \times \tan(60°)$
      = 5437.8kN/m
   μ_2 = coefficient of friction between the treated ground and the foundation
      = 0.6
   $R_2 = W_2' \mu_2 = 5437.8 \times 0.6 = 3262.7$kN/m

$\Sigma H = (H_1 + H_2)$

= (total weight of caisson + total weight of treated ground) $\times$ design seismic coefficient

$= (2505.1 + 4942.2) \times 0.25 = 1861.8 \text{kN/m}$

$$Fa = \frac{R_1 + R_2 + R_3}{\Sigma H + P_h + P_{w2}}$$

$$= \frac{1105.9 + 3262.7 + 455.8}{1861.8 + 2489.1 + 132.9} = 1.08 > Fsa = 1.0$$

5.7.3 The relationship between the slip safety factor and the inclination at the back surface of the treated ground (when the area behind the treated ground has a possibility of liquefaction).

1. *Items studied.* The study is performed using the section shown in Figure 5.29. When the ground behind the treated ground liquefies, the purpose of this study is to investigate how far the slip safety factor varies with the angle (ψ) made by the vertical face and the treated ground face. To examine the slip safety factor, ψ was made to change variously. But, even if ψ changed, the improvement volume of the treated ground was adjusted to make it constant for the examination.

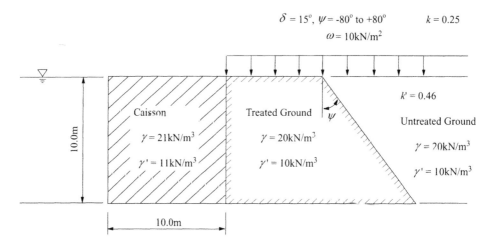

Figure 5.29 Section for study.

2. *Examination result.* Figure 5.30 displays examination results. The figure shows that the slip safety factor decreases with ψ. The reason is that hydraulic pressure from the liquefied ground decreases the effective weight of the treated ground, when the ψ is negative, as described in Section 5.2.4. Therefore, the ψ is desirable to be positive.

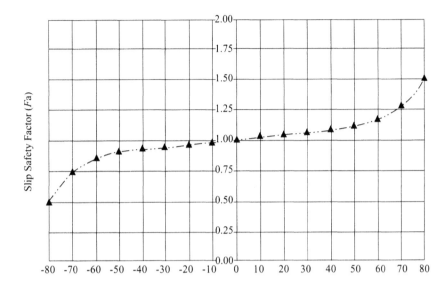

Angle of Treated Ground Face and Vertical Face (ψ)

Figure 5.30 Relationship between slip safety factor (*F*s) and angle between treated ground face and vertical face (ψ).

References

Arai H., Yokoi T., 1965. Study of the seismic resistance of sheet pile walls report No. 3. *Proc. of the 3rd conference of the port and harbor research institute*: 103 (in Japanese).

Coastal Development Institute of Technology, 1997. *Handbook on liquefaction remediation of reclaimed land*, revised edition: 230-238, (in Japanese).

Coastal Development Institute of Technology, 1997. *Handbook on liquefaction remediation of reclaimed land,* revised edition: 78-109 (in Japanese).

Coastal Development Institute of Technology, 1989. *Design of improved ground by the premixing treatment method*: 1-51 (in Japanese).

Japan Port and Harbor Association, 1999. *Technical standards and commentaries for port and harbour facilities in Japan*: 507-510 (in Japanese).

Japan Road Association, 1996. *Specifications for highway bridges*, Part V, seismic design: 53-54 (in Japanese).

Nozu A., Uwabe T., et. al., 1997. Relation between seismic coefficient and peak ground acceleration estimated from attenuation relations. *Technical note of the port and harbour research institute*, No. 893 (in Japanese).

Okabe S., 1924. *General theory on earth pressure and seismic stability of retaining wall an dam*, JSCE, Vol. 10, No. 6 (in Japanese).

The Japan Port and Harbour Association, 1999. *Technical standards and commentaries for port and harbour facilities in Japan*: 207-210 (in Japanese).

The Japan Port and Harbour Association, 1999. *Technical standards and commentaries for port and harbour facilities in Japan*: 254-259 (in Japanese).

The Soil Stable Material Committees Edition, 1991. *The Society of Materials Science, Japan: Soil Improvement.* THE NIKKAN KOGYO SHINBUN. LTD. (in Japanese).

Tuchida T., et. al., 1996. *Calculation of earth pressure during earthquake by slice method, proceedings of thirty-first Japan National Conference on geotechnical engineering.* The Japanese Geotechnical Society: 1083-1084 (in Japanese).

Zen K., Yamazaki H., et. al., 1992. Dynamic strength and deformation characteristics of a cement-treated sand. *The 27th Japan National Conference on soil mechanics and foundation engineering.* The Japanese society of soil mechanics and foundation engineering: 933-934 (in Japanese).

Zen K., Yamazaki H., et. al., 1990. Strength and deformation characteristics of cement treated sands used for premixing method. *Report of the port and habour reserch institute*, Vol. 29, No. 2: 85-118 (in Japanese).

CHAPTER 6

Installation Method and Inspection in Construction Work

6.1 INSTALLATION WORK AND FACILITY

The procedure for the installation is :
  1) transporting landfill material,
  2) mixing the cement,
  3) adding the separation inhibitor,
  4) transporting the treated soil, and
  5) reclamation.
The selection of a suitable combination of the methods is made based on the conditions at the construction site.

6.1.1 Categories of installation methods

Table 6.1 shows mixing methods, mixing locations and reclamation methods. Suitable installation methods vary according to the conditions at each site. Consequently, the past use of installation methods must be considered during the selection process.

Table 6.1 Installation work.

Method of installation	Installation content
Mixing method	Mixing with belt conveyor Mixing with mechanical mixer Mixing with large equipment
Mixing location	Mixing on land Mixing in open water
Reclamation method	Placement using bulldozers Method using a chute Method using open bottom type barges Method using a clam shell Method using open bottom buckets

6.1.2 *Mixing work*

The following mixing methods have been used with the premixing method:
- Belt conveyor mixing method with dumper chutes installed on the connection with each belt conveyor
- Mechanical mixing using a screw mixer or a rotary drum type mixer
- Heavy machine method using excavation equipment such as a backhoe

1. *Belt conveyor mixing method (dry method).* This mixing method is economically and efficiently applied when the landfill material is dry and the water content is less than the natural water content. As shown in Figure 6.1, the facility is composed of a series of items of equipment from the sand hopper to the belt conveyor and separation inhibitor spray nozzle.

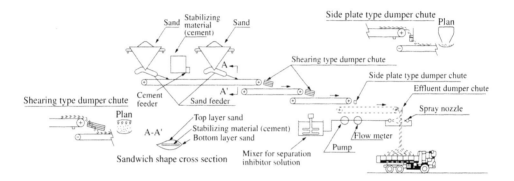

Figure 6.1 Schematic diagram of belt conveyor mixing method.

The characteristics of the facility are as follows:
1) The sand is thinly spread from two sand hoppers on the belt conveyor.
2) The cement is sandwiched between the thin layers of sand on the belt conveyor.
3) Two types of dumper chutes, the shearing type dumper chute (Photo 6.1) and side plate type dumper chute (Photo 6.2), are provided at the connection to each belt conveyor.

The advantages of this system are as follows:
1) Effective mixing is possible with minimal mixing energy because sand is spread thinly during the initial stage of mixing.
2) Dispersion of the cement is prevented by sandwiching.
3) A large volume can be mixed by running the belt conveyor continuously.

Because of these advantages, efficient mixing equal to mechanical mixing is attained. The separation inhibitor is sprayed onto the treated soil as it is discharged from the belt conveyor. Mixing may be done on land or sea according to the site conditions.

It is best to install the mixing plant as close as possible to the landfill site because of the time required for solidification of the treated soil.[2] When a landfill site is close to the loading location for the fill material, it is possible to incorporate the mixing plant in the loading facility. In mixing at sea, a plant may be installed on the barge. A scheme for the mixing plant vessel is shown in Figure 6.2.

Photo 6.1 Shearing type dumper chute.   Photo 6.2 Side plate type dumper chute.

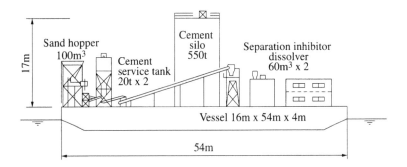

Figure 6.2 Schematic diagram of mixing plant vessel of a mixing plant barge.

2. *Mechanical mixing method (wet method).* The belt conveyor mixing method is difficult to carry out when the water content of the landfill material is relatively high. In such cases, a forced 2-shaft mixer or other mechanical mixer is used. A scheme for this method is shown in Figure 6.3. After mixing, the treated soil is transported by a belt conveyor, which is either placed at the stipulated place as it is, or else placed in a bucket.

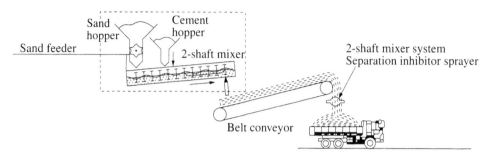

Figure 6.3 Schematic diagram of the forced 2-shaft mixer method.

3. *Heavy machine mixing method.* This mixing method using a backhoe or a stabilizer is a relatively easy method performed using simple machinery. However, the degree of mixing may be uneven, so this method must be planned carefully.

4. *Others.* Mixing methods such as the in-pipe mixing method that are now being developed can be applied after their mixing efficiency and the degree of mixing has been confirmed.

6.1.3 *Mixing sites*

The sites where the stabilizer is mixed are follows: using a sand loading facility at the borrow-pit (mixing on land), using a mixing plant barge floating on the sea (mixing in open water), and installing a mixing facility at the landfill site.

1. *Mixing on land.* The land mixing method is advantageous in cost because this method can be used by adding stabilizer supply equipment to the existing facility.

There is a danger, however, of its strength declining if it takes 2 hours or more to transport, because the treated soil begins to solidify immediately after it is mixed. Therefore, this method must be planned carefully.

2. *Mixing in open water.* The benefits of the sea mixing method are that there is no danger of the strength declining and any kind of sand loading plant can be used, because this method allows the material to be placed immediately after it has been mixed. However, this method is a little more costly than the land mixing method because new mixing equipment must be obtained.

3. *Other mixing methods.* The method of installing the mixing facility at a landfill site can be applied when there is sufficient space for a stock pile of sand and the mixing facility to be installed on the site. The benefits of this method are that it permits the material to be placed immediately after it has been mixed and that any kind of sand loading plant can be used. It is a little more costly than other methods because a sand loading facility and a new mixing facility are necessary.

6.1.4 *Reclamation method*

Several methods can be applied for reclamation using treated soil.

(1) *Spreading method*. This method is shown in Figure 6.4 (a).

(2) *Chute method*. This method is shown in Figure 6.4 (b).

(3) *Direct placement method with bottom dump type barges*. This method is shown in Figure 6.4 (c).

(4) *Clamshell method*. This method is shown in Figure 6.4 (d).

(5) *Bottom dump type bucket method*. This method is shown in Figure 6.4 (e).

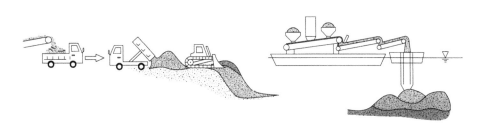

(a) Placement using bulldozers   (b) Method using a chute

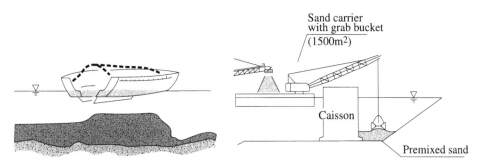

(c) Method using open bottom type barge   (d) Method using a clam shell

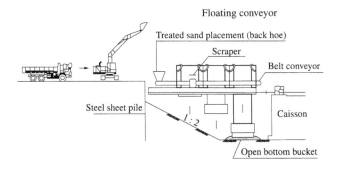

(e) Method using open bottom buckets

Figure 6.4 Landfill site.

Method [1] is suitable for reclamation from land with equipment such as a bulldozer and a backhoe; this method is applied for reclamation in shallow or narrow places where it is not possible to directly place the material because of the draft of barges.

Method [2] is applied at deep locations; this method effectively prevents pollution of the water because the bottom end of the chute is constantly close to the surface of the ground.

Method [3] is suitable for reclamation within an open water area enclosed by a bulkhead; this method is the same as the conventional method presently used.

Methods [4] and [5] are effective with material containing gravel when separation would be a serious problem.

When placing where the water is very deep, continuous execution must be avoided and a staged execution must be performed, as shown in Figure 6.5.

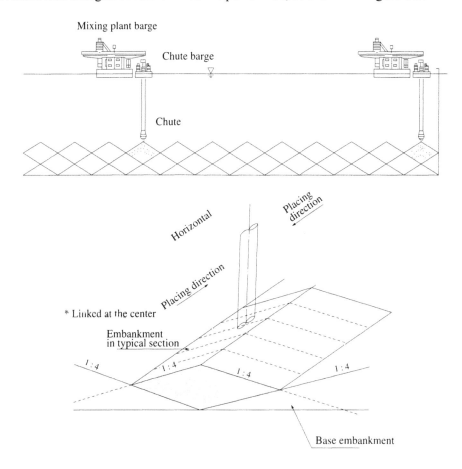

Figure 6.5 Landfill site.

6.2 INSPECTION IN CONSTRUCTION WORK

Inspection consists of four items: inspection of quality, inspection of water quality, inspection of water temperature, quality checks.

6.2.1 *Inspection of quality*
Inspection of quality includes inspection of material, inspection during mixing, and inspection during reclamation.

1. *Inspection of material*. Inspection of material is done for the three materials of sand, stabilizer, and separation inhibitor, to guarantee the mixing properties at the mixing plant. For sand, grain size distribution, density of grains, and water content are inspected. When sand containing gravel is used, it is particularly important to control the maximum grain size because mixing plant functions might be damaged.

The stabilizer and the separation inhibitor are controlled based on the results of inspections when the material is shipped.

2. *Inspection during mixing*. The strength of the reclaimed land depends not only on the percentage of cement added but also on the density. For land reclamation, however, density control is not usually done and hence the percentage of cement added must be strictly controlled in order to acquire the designated strength. The items to be measured are shown in Table 6.2. The measurement is made with a sand feeder, cement feeder, and separation inhibitor pump (variable pump) installed on the mixing plant, as shown in Figures 6.1 and 6.3.

Table 6.2 Measurement in mixing.

Material	Measurement	Control
Sand	Conveyor scale	Sand feeder
Cement	Impact flow meter (belt conveyor mixing) Load cell (mechanical mixing)	Cement feeder
Separation inhibitor	Flow meter	Separation inhibitor pump (variable pump)

As the mix proportion of the stabilizer and the separation inhibitor directly influence the strength of the treated ground, adequate control is achieved and they are measured accurately and supplied so as to obtain a fixed quality. The percentages of the stabilizer and the separation inhibitor added are set relative to the dry weight of the landfill soil. The following are examples of the measurement of landfill soil, stabilizer, and separation inhibitor when the material is mixed by a belt conveyor on the sea.

2.1 *Quantity of landfill soil supplied*. The quantity of landfill soil supplied is continuously measured by the conveyer scale; the sand feeder is controlled automatically by the signal from this conveyer scale, and in this way, the quantity supplied is always maintained at the set value. The cumulative quantity of sand is automatically recorded and controlled.

2.2 *Quantity of stabilizer supplied.* The weight of the stabilizer supplied is continuously recorded by a stabilizer scale; the stabilizer feeder is controlled automatically by the signal from this scale, and in this way, the quantity supplied is always maintained at the set value. The cumulative weight of the stabilizer is automatically recorded and controlled.

2.3 *Quantity of separation inhibitor added.* The separation inhibitor added by the spray in solution is measured by a flow meter; the separation inhibitor pump is controlled automatically by the signal from this flow meter; and in this way, the quantity supplied is always maintained at the set value. The cumulative quantity of separation inhibitor is automatically recorded and controlled.

2.4 *Sand water content.* The quantity of sand supplied changes according to the water content of the sand. Therefore, the water content of the soil is measured. The measurement of the water content of the sand is performed continuously by a non-contact sensor on the belt conveyor (before mixing), and the adjustment of the quantity of sand supplied is performed as necessary. The water content of pit sand or sand from a stock pile is almost constant. Table 6.3 shows the results of this measurement. Figure 6.6 shows a schematic diagram of measurement in the sea mixing method.

Table 6.3 Volume change rate, unit weight, and water content of ground.

Item	Ground	Stock pile
Volume change rate	1.00	1.30
Wet unit weight (kN/m^3)	16.2	12.5
Dry unit weight (kN/m^3)	17.8	13.7
Water content (%)	10.0	10.0

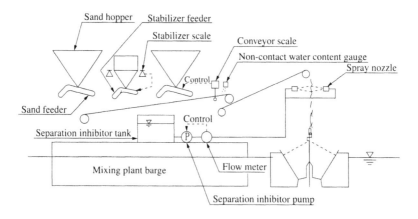

Figure 6.6 A schematic diagram of measurement.

Samples of mixed sand discharged from the belt conveyor are obtained in order to perform the calcium analysis and unconfined compression test.

3. *Inspection during reclamation.* When treated soil is deposited underwater, the slope of the mound thus placed depends on the grain size of the fill material. In fine sand, the inclined slope is gentle but in coarse grain sand, it is steeper. When reclaiming the treated soil by chute, the slope inclination becomes about 1:4.

Generally, the inclination depends on the method of reclamation and type of soil. It is necessary to investigate the shape of the mound with a preliminary field test before construction. When reclaiming with a chute, the mound shape can be obtained by measuring the water depth at the perimeter of the mound with an automatic sounding device. The shape of the entire mound created is checked with a lead sounding device or bathometer. Check sampling during reclamation is performed by mold sampling.

6.2.2 *Inspection of water quality*

1. *Environmental surveys and tests.* Environmental surveys and tests must be performed correctly before, during and after construction, and the influence on the surrounding environment must be confirmed by comparing the results of each of the surveys and tests. Figure 6.7 shows a flow chart of the environmental conservation measures done before and during construction.

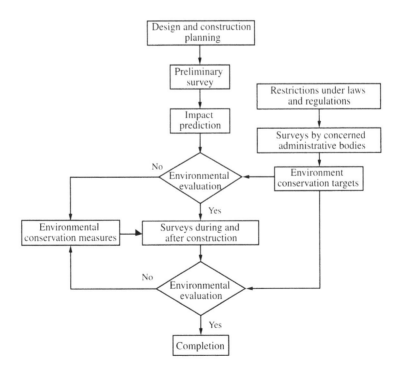

Figure 6.7 Flow chart of the environmental conservation measures.

The premixing method produces far less noise and vibration than the sand compaction pile method and similar methods, but it is necessary to be careful about water pollution. It is, therefore, extremely important to observe laws, regulations, and standards governing water pollution. In addition, it is necessary to survey and study other relevant items.

The following are the items about water pollution that are surveyed and studied as necessary.

1) pH (hydrogen ion concentration)
2) SS (suspended solids)
3) Turbidity
4) DO (dissolved oxygen)
5) COD (chemical oxygen demand)
6) Clarity

Examination of sea bottom topography, depth, water temperature, tidal current, and tide level in relation to the water quality investigation contributes greatly to survey analysis. There is a local government organization which regularly carries out water quality investigations, so a method of using the results has been established.

2. *Inspection of water quality.* The separation inhibitor forms the treated soil into lumps and at the same time suppresses the separation of the cement and promotes settlement of suspended soil particles. Therefore, diffusion of turbidity and elution of alkali are suppressed.

Within the water surrounding a treated soil landfill site, the pH is high. However, the seawater has a neutralizing effect so that the eluted alkali is diluted by the seawater and the pH is lowered, although this is not a serious problem.

Water quality inspection during installation is executed with a focus on the turbidity and pH at representative locations inside and outside the installation area, inside and outside the turbidity prevention cloth, and near the fill placement. Figures 6.8 to 6.11 show examples of measured results of water quality in construction.

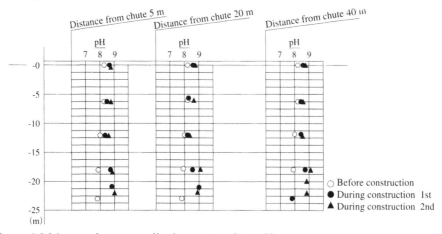

Figure 6.8 Measured water quality in construction (pH).

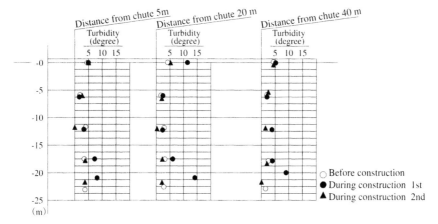

Figure 6.9 Measured water quality in construction (Turbidity).

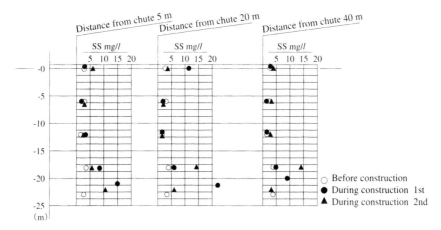

Figure 6.10 Measured water quality in construction (SS).

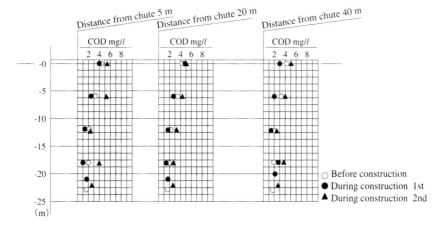

Figure 6.11 Measured water quality in construction (COD).

3. *Changing water quality*. Environmental conservation measures suited for each construction site must be implemented, because the changes in water quality caused by the premixing method vary according to the installation method and the construction site. With this method, cement and separation inhibitors made from water-soluble polymer are added to the treated soil as the stabilizer, so their influence must be studied in advance. The following are the results of studies of changes in water quality caused by the premixing method.

3.1 *Studies of changes in water quality*.

3.1.1 *Stabilizer concentration and pH*. Figure 6.12 shows the results of a study of the relationship of stabilizer concentration with pH values, when the stabilizer is added to and spread in sea water and in city water.

The pH value rose as the stabilizer concentration increased, and remained almost constant at 10.1 once the stabilizer concentration exceeded 1,000 mg/liter in the sea water.

In the city water, the pH value changed sharply according to the stabilizer concentration; for example, the pH value rose to about 11.5 at a stabilizer concentration of 1,000 mg/liter and reached 12 at a stabilizer concentration of 10,000 mg/liter.

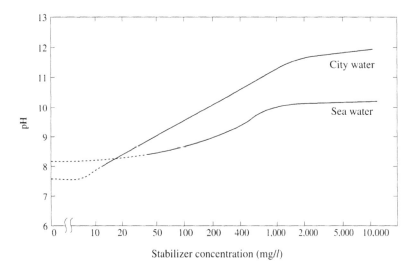

Figure 6.12 Stabilizer concentration – pH relationship.

3.1.2 *Reduction of the pH value by dilution*. Figure 6.13 shows the results of a study of the pH value, when the alkali elution that is obtained by filtering the stabilizer which remained 24 hours after adding the saturating stabilizer to the sea water, is diluted by various multipliers of sea water. The pH value fell sharply due to the dilution, and the influence of the alkali almost disappeared when the elution was diluted 50 times.

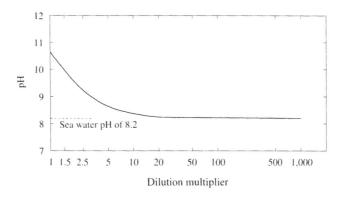

Figure 6.13 Reduction of pH by sea water dilution of alkali elution.

3.1.3 *Diffusion*. Changes in the pH value and turbidity of the water were measured, when the treated soil shown in Table 6.4 was placed from a bottom dump type barge constructed to a scale of 1/7. This experiment was performed with a large outdoor tank (16 m×6 m×3.3 m) filled with city water to a depth of 2.57 m. Figure 6.14 shows the way the measurements were made, and Figures 6.15 and 6.16 show the results of the measurements.

Table 6.4 Treated soil mix proportions (large outdoor water tank).

Landfill material	Rokko masa soil (Maximum grain size : 38.1 mm)
Stabilizer	Ordinary portland cement: 4%
Separation inhibitor	Polyacrylamide: 0 mg/kg or 200 mg/kg
Mixing method	Mixing using a belt conveyor

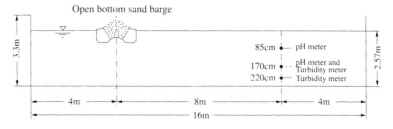

Figure 6.14 Measurement method (large outdoor water tank).

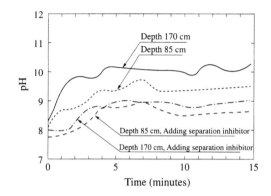

Figure 6.15 Results of pH measurement (large outdoor water tank).

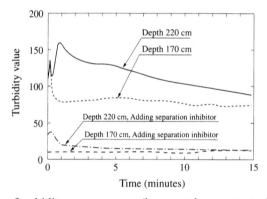

Figure 6.16 Results of turbidity measurement (large outdoor water tank).

Changes in the pH value were measured, when the treated soil shown in Table 6.4 was continuously placed at a rate of 300g every two minutes. The next experiment was performed with an indoor tank (1 m×1 m×0.4 m) filled with sea water to a depth of 0.2 m. Table 6.5 shows the mix proportion of the treated soil, Figure 6.17 shows the measurement method, and Figures 6.18 and 6.19 show the results of the measurements.

Table 6.5. Treated soil mix proportions (indoor water tank).

Landfill material	Rokko masa soil (Maximum grain size : 5.74 mm)
Stabilizer	Slag cement (5%)
Separation inhibitor	Not used
Turbidity prevention cloth	Coefficient of permeability 1.5×10^{-2} cm/s

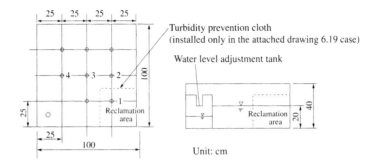

Figure 6.17 Measurement method (indoor water tank).

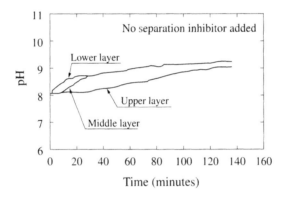

Figure 6.18 Diffusion of pH in the depth direction (indoor water tank).

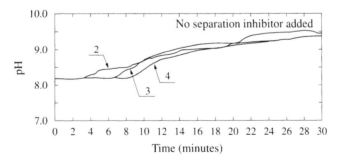

Figure 6.19 Diffusion of pH in the horizontal direction (indoor water tank).

In every case, the pH value was high at measurement points where the water was deep. At measurement points in the upper layer, the effects appeared later than in the lower layer. In particular, as shown in Figure 6.18, the vertical flow of water was little and the pH value rose quickly in the lower layer but rose slowly in the upper layer, when the quantity of placed sand was small relative to the size of the water tank.

Regarding the horizontal diffusion, the difference of the pH value appeared according to the location until the fifth placement of sand was performed (after

10 minutes), but the difference did not appear after that. The turbidity fell with time, but the pH value did not. The pH and turbidity were lower when the separation inhibitor was added than when it was not added, as shown in Figures 6.15 and 6.16.

Changes in the pH value and the turbidity were measured when the treated soil shown in Table 6.6 was placed (the placing rate was 80 m^3/hour) using the turbidity prevention type chute in the large outdoor tank (6.74 m × 6.74 m × 10 m (water depth). Figure 6.20 shows the water quality survey points. Figures 6.23 and 6.24 show the relationship of the value of pH and turbidity and the increase in the elevation of the tip of the chute, when the vertical axis shows the value of pH and turbidity and the horizontal axis shows the increase in the elevation of the tip of the chute. The height of tip of the chute corresponds to the reclamation progress.

Table 6.6 Mix proportion of the treated soil (large water tank placing experiment, placed at 80 m^3/hour).

Landfill material	Kinamidayama sand in Chiba Prefecture
	(Maximum grain size 4.76 mm)
Stabilizer	Portland blast furnace slag cement Type B: 7.5%
Separation inhibitor	Strong anion type polyacryldmide: 30 mg/kg
Mixing method	Mixed using a belt conveyor

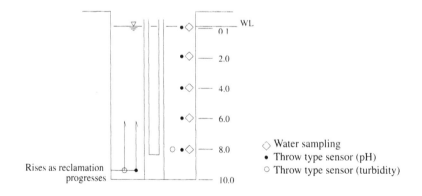

Figure 6.20 Water quality survey points.

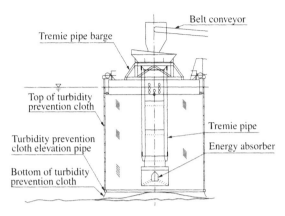

Figure 6.21 Turbidity prevention type chute.

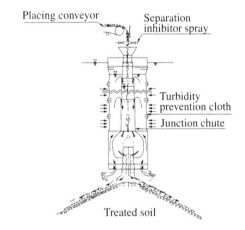

Figure 6.22 Turbidity prevention type junction chute.

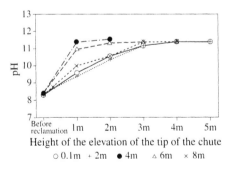

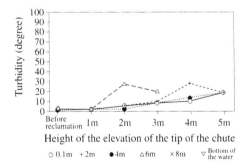

Figure 6.23 pH measurement results.   Figure 6.24 Turbidity measurement results.

Changes in the pH value and the turbidity were measured when the treated soil shown in Table 6.7 was placed (the placing rate was 250 m³/hour) using the turbidity prevention type junction chute in the large outdoor tank (6.74 m × 6.74

m×10 m (water depth). Figure 6.25 shows the measurement points, and Figures 6.26 and 6.27 show measurement results.

Table 6.7 Mix proportion of the treated soil (large water tank placing experiment, placed at 250 m³/hour).

Landfill material	Kinamidayama sand in Chiba Prefecture
	(Maximum grain size 4.76 mm)
Stabilizer	Portland blast furnace slag cement Type B: 7.5%
Separation inhibitor	Strong anion type polyacryldmide: 30 mg/kg
Mixing method	Mixed using a belt conveyor

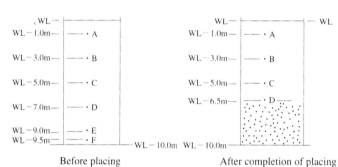

Figure 6.25 Water quality survey points.

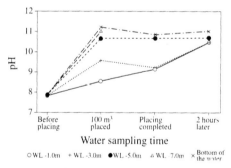

Figure 6.26 pH measurement results.

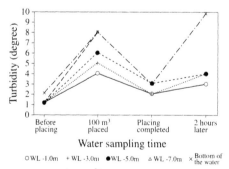

Figure 6.27 Turbidity measurement results.

These two water quality studies based on large water tank experiments have revealed the following facts.

Ұ The pH value rises rapidly after embankment construction, but, after that, it rises slowly and remains constant at about 11.

Ұ When compared by depth, the pH value rises rapidly as the measurement point becomes deeper.

Ұ The turbidity rises only slightly, and the value of the turbidity does not reach 10 even after reclamation with the chute has been completed.
After reclamation is restarted, at first the turbidity does rise, but as the reclamation advances to a certain degree, the turbidity tends to fall.

Ұ The pH value showed the same tendencies when the soil was placed at 80 m³/hour and at 250 m³/hour.

However, the turbidity was lower in the latter case, and the experiments have also shown that the junction chute method is an underwater reclamation method that more effectively prevents water pollution.

3.1.4 *Prevention of diffusion by a turbidity prevention cloth.* Changes in the pH value were measured, when the treated soil shown in Table 6.7 was placed continuously up to the water surface enclosed by the turbidity prevention cloth. The measurement was performed in the same way as shown in Figure 6.17, and the measurement results are shown in Figure 6.28. Within the reclamation zone, the pH value rose abruptly after the first stage of the reclamation and that value became a constant value of about 10; but outside the reclamation zone (outside the turbidity prevention cloth), the pH value only rose slightly. No turbidity was observed outside the turbidity prevention cloth.

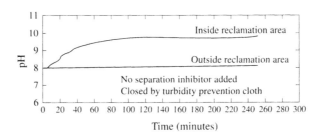

Figure 6.28 Prevention of diffusion of pH by turbidity prevention cloth.

3.1.5 *Water quality survey in open water*

Ұ Changes in water quality caused by direct placing

The water quality was measured, when 132 m³ of the treated soil shown in Table 6.8 was divided into three and placed directly. The sand was placed directly at a depth of about 10 m. Figure 6.29 shows the way the measurements were performed and Figures 6.30 to 6.33 show the measurement results.

Table 6.8 Mix proportion of the treated soil (placed in the ocean).

Landfill material	Iejima gravelly sand (Maximum grain size 50 mm)
Stabilizer	Slag cement: 5.5%
Separation inhibitor	Polyacrylamide: 50 mg/kg
Mixing method	Mixed using a concrete mixing barge

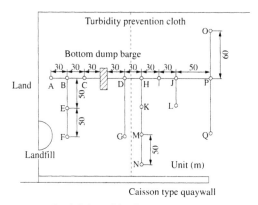

Figure 6.29 Measurement method (placed in the ocean).

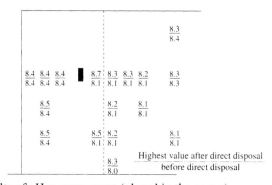

Figure 6.30 Results of pH measurement (placed in the ocean).

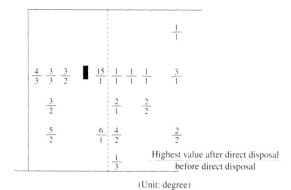

Figure 6.31 Results of turbidity measurement (placed in the ocean).

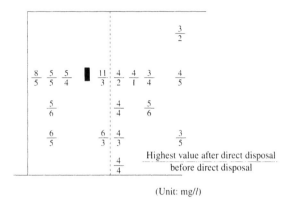

(Unit: mg/*l*)

Figure 6.32 Results of SS measurement (placed in the ocean).

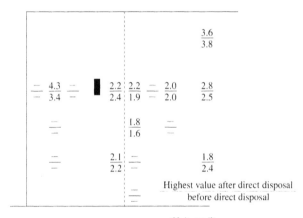

(Unit: mg/*l*)

Figure 6.33 Results of COD measurement (placed in the ocean).

Regarding any changes in water quality caused by direct placing, the value of the pH, turbidity, and SS rose at the deep point, but no changes in the COD value appeared. The turbidity fell as time passed after the direct placing. Outside the turbidity prevention cloth, no changes in any measured qualities were observed at any measurement points. Changes in water quality caused by reclamation by backhoe.

The water quality was also measured, when 300 m^3 of the treated soil shown in Table 6.9 was placed by reclamation using a backhoe. Figure 6.34 shows the way the measurements were performed and Figures 6.35 to 6.37 show the measurement results.

Table 6.9 Mix proportion of the treated soil (reclamation by backhoe in the ocean).

Landfill material	Tokushima gravelly sand (Maximum grain size 300 mm to 400 mm)
Stabilizer	Slag cement: 4.0%
Separation inhibitor	Not used
Mixing method	Mixed using a back hoe (5m³/min)

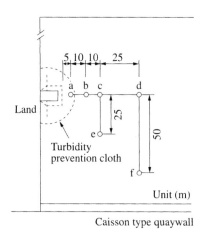

Figure 6.34 Measurement method (reclamation by backhoe in the ocean).

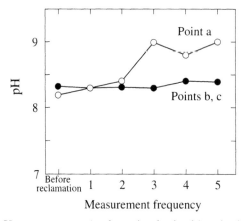

Figure 6.35 Results of pH measurement (reclamation by backhoe in the ocean).

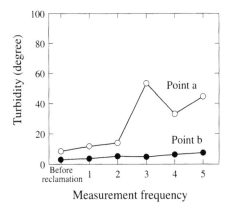

Figure 6.36 Results of turbidity measurement (reclamation by backhoe in the ocean).

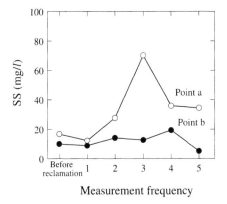

Figure 6.37 Results of SS measurement (reclamation by backhoe in the ocean).

No change in COD values were observed at any measurement point. At points not shown in the figure, no changes were observed in the pH, turbidity, and SS values. Within the turbidity prevention cloth, the values of these rose after the start of construction, but, outside, the values of the pH and turbidity rose slightly after the construction was far advanced.

3.1.6 *Neutralization of alkali.* Changes of the pH value were measured, when the stabilizer was suspended in 480 liters of city water and carbon dioxide was pumped into this water. Figure 6.38 shows the results of these measurements. The pH value was lowered as the gas was pumped, and the higher the stabilizer concentration, the more gas was required to lower the pH value.

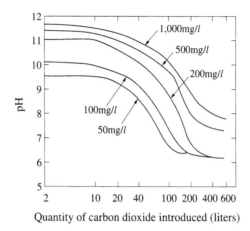

Figure 6.38 Reduction of pH by carbon dioxide.

3.2 *Summary.* Regarding the water quality in the premixing method, changes of the value of pH, SS, and turbidity appeared, but the value of COD did not change. However, when the sand was placed directly, the value of COD did rise when the separation inhibitor concentration in sea water exceeded 25 mg/liter. For reclamation using the premixing method, as a way of preventing water pollution, the same methods regarding SS and turbidity used in conventional construction can be implemented.

The following three points prevent the pH from rising.

3.2.1 The separation and diffusion of the stabilizer must be prevented.
3.2.2 The dilution of alkali by water is enough.
3.2.3 Neutralization is caused by acid material.

Regarding point [1], the use of a separation inhibitor and the installation of stand-alone turbidity prevention clothes on the seabed are effective.

6.2.3 *Inspection of water temperature*

It is necessary to be extremely careful in the inspection with quality with the water temperature, because the strength of the treated soil is strongly influenced by the sea water temperature, as described in Chapter 2. When the water temperature is particularly low, an initial strength development has a tendency to become late with any decrease in the percentage of cement added.

6.2.4 *Inspection of quality of treated soil/fill*

A post-execution check boring is performed to confirm the quality of the treated soil and fill, as shown in Table 6.10.

6.3 PRELIMINARY FIELD TEST BEFORE CONSTRUCTION

A preliminary field test is recommended before construction if there is no construction experience at the site. The test is performed with a prearranged mixing plant, equipment, and reclamation procedure, using a number of mix proportions previously set based on laboratory mix proportion tests for the treated soil. Verification is made that the required strength is obtained. In addition, the field test is designed to clarify in detail and to confirm the mix proportion at the job site, construction, inspection methods, and yield.

The major items surveyed and tested are shown in Table 6.10.

Table 6.10 Survey and test in preliminary field test.

Item	Purpose	Method
Testing properties of the landfill material	Determining that the material used, particularly the soil is suited to the construction	Density of the soil particle, grain size, water content, maximum and minimum density of the soil
		Results of properties testing of the stabilizer and separation inhibitor
Test of the measurement of the material	Determining that the measurement of the material used can be performed correctly	Inspection of the weight of the material weighed by the planned measurement
Survey of the percentage of stabilizer added	Examination of the cement content and strength of the treated sand and scattering of these values to determine the required additional rate and decide if the percentage of stabilizer added is appropriate to the design strength	Calcium analysis of the treated sand
		Unconfined compression test and density test of the treated soil
		Unconfined compression test and density test of the treated sand
Survey of the completed reclamation work quality	Post-construction examination of the form of landfill and its completed work quality in order to determine if the placing and reclamation method, and completed work quality control method are appropriate	Measurement of the depth of the landfill
Survey and testing of the treated fill	Examination of the strength and properties of the treated fill in order to determine if the design strength can be achieved	In-situ survey (standard penetration test, elastic wave exploration) Unconfined compression test Triaxial compression test Cyclic undrained triaxial test

References

Katano H., Kuroyama H., et al., 1987. Research on Landfill Methods Using Improved Sandy Material - Belt conveyor Mixing Test. *Proc. of the 42nd Annual Conference of the Japan Society of Civil Engineers*, 3: 826-827 (in Japanese).

Kuroyama H., Tamai A., et al., 1987. Research on Landfill Methods Using Improved Sandy Material (Part 4) – Effects on Water Quality. *Proc. of the Fourteenth Annual Conference of the Kanto Chapter of the Japan Society of Civil Engineers*: 164-165 (in Japanese).

Mori T., Kubo H., et al., 1991. Development of a Premixing Treatment Method, Reclamation Test by the Chute Method, Part 5 (Mechanical Properties of Young Treated Soil), *Proc. of the 46th Annual Conference of the Japan Society of Civil Engineers*, 3: 1038-1039 (in Japanese).

For Product Safety Concerns and Information please contact our EU
representative GPSR@taylorandfrancis.com
Taylor & Francis Verlag GmbH, Kaufingerstraße 24, 80331 München, Germany

ⁿcontent.com/pod-product-compliance
ⁿnt Group UK Ltd.
Keynes, MK11 3LW, UK
ⁿ0425
97B/236